Quarterly Essay

Quarterly Essay is published four times a year by Black Inc., an imprint of Schwartz Publishing Pty Ltd. Publisher: Morry Schwartz.

ISBN 978-1-86395-921-6 ISSN 1832-0953

Subscriptions – 1 year print & digital (4 issues): $79.95 within Australia incl. GST. Outside Australia $119.95. 2 years print & digital (8 issues): $149.95 within Australia incl. GST. 1 year digital only: $49.95.

Payment may be made by Mastercard or Visa, or by cheque made out to Schwartz Publishing. Payment includes postage and handling.

To subscribe, fill out and post the subscription card or form inside this issue, or subscribe online:

www.quarterlyessay.com
subscribe@blackincbooks.com
Phone: 61 3 9486 0288

Correspondence should be addressed to:

The Editor, Quarterly Essay
Level 1, 221 Drummond Street
Carlton VIC 3053 Australia
Phone: 61 3 9486 0288 / Fax: 61 3 9011 6106
Email: quarterlyessay@blackincbooks.com

Editor: Chris Feik. Management: Caitlin Yates. Publicity: Anna Lensky. Design: Guy Mirabella. Assistant Editor: Kirstie Innes-Will. Production Coordinator: Hanako Smith. Typesetting: Tristan Main.

THE LONG GOODBYE

Coal, Coral and Australia's Climate Deadlock

Anna Krien

I had no expectations. If anything, I was anticipating disappointment. Not so much because of premature obituaries of the Great Barrier Reef; it was more my distrust of adjectives. I have been to enough bucket-list tourist sites to know the experience can leave one feeling strangely empty. I sat next to the skipper, blue water whooshing underneath us, while, on the deck below, German backpackers organised their underwater cameras and diving gear. A few of us watched the shoreline, Townsville shrinking as we got further away. Castle Hill, a mass of orange granite that separates the city centre from the urban sprawl beyond, glowed in the pink morning light. Rusted to the landscape is Clive Palmer's Yabulu nickel refinery, its three smokestacks no longer puffing. There's a lush strand I had walked that morning on my way to the jetty, where black cockatoos hang upside down, tearing fruit off the trees as early morning joggers dodge the leaf litter. I passed a brightly painted water playground, its movement sensors waiting for the day's first chubby bather-clad preschooler.

It's often said in news reports that you can see the Great Barrier Reef, the world's largest structure made up of living organisms, from space – as if that will convince us to protect it. But you can see a lot of things from space: China's smog, and the "Super Pit," a goldmine in Kalgoorlie.

In 1968, on the first human voyage to orbit the moon, the American astronaut Bill Anders was given the job of photographer. The *Apollo 8* mission was focused on the moon, but then . . .

> ANDERS: Oh, my God! Look at that picture over there! There's the Earth coming up. Wow, is that pretty.
> BORMAN: Hey, don't take that, it's not scheduled. (*joking*)
> ANDERS: (*laughs*) You got a colour film, Jim? Hand me that roll of colour quick, would you . . .
> LOVELL: Oh man, that's great!

Anders' words are often repeated: "We came all this way to explore the moon, and the most important thing is that we discovered the Earth." His photo, titled *Earthrise*, became a lightning rod for environmentalists, who knew the blue and green swirling globe coming out of the dark was a gift too precious to lose sight of. It is a difficult task, to remember we are on a cat's-eye marble in a pitch-black void.

"Are we in space?" my four-year-old had asked me in Melbourne, not long before I left for my reef visit.

"No," I replied confidently, then just as quickly lost confidence. "Well, actually, we are. We are on Earth, and Earth is kind of wrapped, like a present [his eyes lit up], in our atmosphere, the sky, and all of this is in space. So we kind of are in space."

He nodded solemnly, absorbing this before continuing with his painting, but I was trembling. It was that strange feeling of having known something for a long time, but the act of saying it out loud making it real. *We are in space.* I wanted to scream it on the street like a crazy woman. We are in space!

Two hundred years of science and discovery have seen a rapid reorientation of many assumptions. The first dizzying insight was that of extinction, proved by paleontologist Georges Cuvier in 1796. It had been assumed until then that life just went on. Sure, it was punctuated by death, mostly in dribs and drabs, sometimes in tragic sweeps like the Black Death, but a species did not simply cease to exist. There would always be another specimen in its mirror image, and another, and another.

It got worse. Cuvier, discovering an increasing diversity of fossils, suggested there had been catastrophes on Earth, many of them. "Life on Earth has often been disturbed by terrible events" – death en masse, in a cataclysmic blink of an eye. Then, in 1859, came the theory of evolution by natural selection. It was considered highly unflattering – we came from apes! – but survival of the fittest was also a balm to Cuvier's catastrophes. Extinction was a by-product of evolution, a race which humans were obviously winning. But the discoveries kept coming. As geologists scaled and studied the strata of the earth, clue upon clue began to reveal a series of rapid extinctions, in which even "fit" species had not had time to adapt. As David Raup, a paleontologist, explained in Elizabeth Kolbert's seminal book The Sixth Extinction, the history of life on Earth consists of "long periods of boredom interrupted occasionally by panic."

Many of these revelations are due, in large part, to the Industrial Revolution – sometimes quite literally, as in the case of Nýřany, a small mining town in the Czech Republic. Here, in 1870, a diverse collection of fossils was discovered by the director of the local natural museum of history, who had gotten into the habit of splitting the lumps of coal servants delivered to his study for heating. Inside, he often found intricate traces of four-legged creatures known as tetrapods. That he made these discoveries as he poured coal into his heater – burning the dark compressed chunks of Permian plant matter, which in turn shapeshifted again, this time into a slender length of smoke up his flue and into the sky, a culmination of atmospheric carbon that one hundred years later would initiate a sixth wave of extinctions – is a modern twist on an old parable. Cause and effect: a simple concept, but with incredibly complex effects.

*

In early 2016 the Great Barrier Reef turned white. Professor Terry Hughes, head of the multi-institutional National Coral Bleaching Taskforce, spent eight days flying over the reef, ranking the coral from unaffected to varying levels of bleached. Sweltering inside the tiny plane, the pilot landed in a cane field so that they could remove the doors. "We left them in the field," Hughes told me. At the same time, the taskforce had 100 researchers underwater. In the entire system – some 2300 kilometres of reefs, atolls and islands stretching from the Torres Strait in the north to Fraser Island in the south – the taskforce found that only 7 per cent of the coral had been completely spared bleaching. In the far north, it was estimated that 80 to 90 per cent of corals were either dead or dying. It usually takes a few months to recover or die from a bleaching event, but this time, in the northern part, much of the coral died instantly. It was fried. Over 65 per cent of the northern reef is dead. In a tweet, Hughes released the survey findings with a message: "I showed the results of aerial surveys of #bleaching on the #GreatBarrierReef to my students, And then we wept."

The following week, in federal parliament, the Tasmanian Greens senator Peter Whish-Wilson read out Hughes' tweet and members of the Coalition laughed and gave sarcastic sighs of sympathy. Meanwhile, further north, vast swathes of green mangroves – 7000 hectares – turned a ghostly white. Off the coast of Western Australia, 100 kilometres of kelp forests were wiped out, and up in the Kimberley much of the coral bleached. Marine scientists returned a few months later and were devastated to find Scott Reef mostly dead and covered in slimy algae. Naysayers of climate change had blamed the east-coast bleaching on an El Niño year – a natural warming cycle that occurs every five years or so – but this should have seen the west coast experience cooler temperatures. It didn't. In Tasmania, the kelp forests along the east coast, once so giant they had to be marked on shipping maps, were pushed a little further into oblivion, while warmer water contributed to massive mortalities at the island state's oyster farms. In Victoria an algal bloom affected up to 1700 kilometres of the Murray River, cutting off towns and farmers from water supplies. Local blooms are common, but extending for 1700 kilometres? This is the new normal.

In February 2017 the Great Barrier Reef bleached once more. Professor Terry Hughes again found himself in a small plane flying over while divers in black wetsuits flopped overboard below.

This time, the middle section of the reef was the worst affected. Welcome to the Myxocene. From the Greek word muxa, meaning "slime," it is what Canadian marine biologist David Pauly has proposed calling the new geological epoch Earth is entering – an epoch uniquely of our own making, a sped-up, tricked-up version of natural warming by an excess of emissions, deforestation, overfishing and pollution.

For a time, until the late 1970s, climate scientists were unable to say with certainty if human-made emissions were going to cause the planet to cool or warm (their formidable task was complicated further by the effect of aerosols on the atmosphere). Today, however, the evidence is unequivocal. Coral reefs are one such indicator. Occurring, with the odd exception, within a band of water from the Tropic of Cancer to the Tropic of Capricorn, these reefs form a ring around the globe. Since the 1980s they have been taking direct hits.

Bleaching was first noted in 1911 at Bird Key Reef in Florida Keys. Since then there has been the odd minor instance, but en masse is a recent phenomenon. It is an irony of sorts that just two years after the Great Barrier Reef was accorded World Heritage status in 1981 – rescuing it from being mined for cheap limestone fertiliser and drilled for oil – the first mass bleaching occurred: this is not a threat that can be controlled by declaring a zone protected.

Global warming is pushing up temperatures in the ocean at a rate too fast for most corals, and all that rely on them, to adapt. An unseen side effect of this temperature rise is a change in chemistry. Carbon and methane emissions are making the sea acidic. A warming and increasingly acidic ocean will see a cascade of effects, such as more algal blooms;

the demise of crustaceans, as their shells become too brittle and difficult to form; and coral reefs turning to rubble, taking with them the livelihoods and sustenance of 500 million people worldwide.

The Myxocene is not in the future; it is already here. In the coastal waters of Japan since the 2000s, plague-like blooms of jellyfish have overwhelmed fishermen, with 500 million or so blotting otherwise empty ocean. Fishing boats began to attach wire grills to the ends of their nets, metal teeth slicing the mass of globs into pieces. It was a bad idea – the gelatinous mincing dispersed billions of eggs from the female jellyfish.

In 2013, in the Pacific Ocean, a bewildering "blob" of warm water parked itself for two years, clinging to North America's west coast. A highly toxic and long-lasting algal bloom saw thousands of sea lions and seabirds, as well as hundreds of otters and whales, suffering seizures and dying, shutting down shellfish farms and other fisheries.

It is expected that weird bodies of mucus in the ocean, known as "dead zones," some over 200 kilometres long, will become more common. In 1991 an Italian marine biologist, Serena Fonda Umani, swam alongside a "mucilage" in the Adriatic Sea. A *National Geographic* article in 2009 described the mass as "too dense to swim inside." The article continued: "She remembers diving about 50 feet (15 meters) down when she got the sensation of a ghost floating over her—'sort of an alien experience.'"

There are people who say this has happened before, often arguing in the same breath that action on climate change is therefore foolish. They are right on the first count. In Precambrian times, before fish evolved, oceans were slimy and hot. As for their conclusion – that nothing need be done to curb emissions – so far, no scientist has discovered evidence of thriving human communities living alongside such oceans.

*

"Like most Australians," Kate Jennings wrote in her essay "An Otter's Life," "I am a swimmer." It was with this one line that I wrestled down doubt about visiting the reef, the feeling there was something gross about my going to see it. I had, after all, felt a hot rage when I saw the *Australian* this year promote its travel supplement with the headline "Endangered Destinations: The things to see before they disappear" – yet here I was, packing my bag. It felt naive too, not so dissimilar from a recent visit to the reef by Pauline Hanson, who flopped into the water in her kit and declared to cameras that the reef is "pristine." What would she know? What would I know? But now, on the boat, passing Magnetic Island, these Melbourne hang-ups were blown away with the wind. It was Sunday. My phone had no reception. Someone was looking after the kids. The boat could stop right here, in the deep blue, plonk me out like a sinker and I'd be content.

Then I saw it. I stood up fast, my arm flung out, pointing. "Is that it?" The skipper nodded. Ahead of us, deep dark water had suddenly transformed into rippling vast strips of turquoise, eggshell blue, a strange luminous green. I felt my chest leap with excitement. Everyone else was below, already dressed in their gear and lining up to fall into the water. I hurried to the ladder and practically slid down to join them. As I did, a flush of electric-blue fish skittered out of the water beside the boat in an arc. I found my snorkel and mask, pulling on the daggy blue stinger suit as the boat's crew leader buddied us up. I was with a girl called Grace, a James Cook University student. We smiled at each other, and then we were in.

In the water, I dived down and closed my eyes, giving my body a twist and roll up to the surface, a little ritual. There was a crackling sound, like the Magic Gum my brothers and I used to buy at the milk bar, a sachet of popping candy you poured onto your tongue, where it fizzed and crackled. Fanned out below was the strangest world of corals, some pulsing as if soft, others brittle, the colours an unexpected autumnal palette of mustard yellow, velvet red, tan suede and brown corduroy. Others pulsed fluorescent purple nipples and tubes of glowing pinks, patterns as intricate as the map of a subway, thousands of tiny tongue-pink tips waving, breathing, and long lumpy sea cucumbers with zebra stripes. In between atolls on a pocket of sand lay a solitary stingray, and the frilly mouths of enormous clams flinched when you wiggled your fingers near them, fluorescent green squiggles glowing and you could see right inside their pearly innards.

"How big were they, Mum?" my four-year-old asked on the phone that evening.

"The clams?"

"Yes!"

"Some were as big as our bathtub, others as big as our kitchen chairs."

"The new kitchen chairs or the old ones?"

"Um, the old ones."

"Wow."

*

Two hours later, on the boat for lunch, the crew leader commended everyone for staying close by, "except for Anna and Grace." With as much sternness as you're allowed to muster towards paying customers, he looked at us: "Were you two just going to keep heading out to sea?" Maybe. At the beginning, we looked up every now and then to check our distance from the boat, but then swam as if in a dream. Now, as we ate sandwiches, the crew leader spoke about the reef. He told us about the crown-of-thorns starfish: native to the reef, it has a voracious appetite, expelling its stomach out of its mouth and over the coral, eating the equivalent of its body mass in one sitting. Forty years ago, populations of the crown-of-thorns starfish began to explode. One of its natural predators, the giant triton, had been extensively collected for its ornamental shell and this was thought to be the cause, but by the '70s it became clear the starfish were also thriving in the nitrogen-rich run-off (mostly fertiliser from sugarcane and banana farms) into the reef's catchments. "They hide under the corals during the day," the crew leader told us, "and come out to eat at night."

After lunch, Grace and I swam out again. There was a single coral that looked like an enormous brain, as big as a VW Beetle. As if my eye was now trained, I saw a crown-of-thorns starfish on top of another coral, busy chomping away, too hungry to wait for night. I tried to dive closer, snorkel in hand, but my ears couldn't take the depth. Grace and I dipped into a small cavern and found ourselves inside a shoal of thousands of electric-blue fish. They shivered around us, then up came a second shoal of slightly bigger turquoise fish and a third shoal of larger green fish with pink threaded scales.

We floated in a trance.

I found that when I waved my hand through the water, the small electric-blue fish would swing away, triggering the next shoal to shift, and so on. I started to move my body like a conductor, using tiny movements to press against and shimmer the shoals, bringing them back, then away, rounding in on them, and pausing so that they would flood back up from the depths. At first we didn't notice the corals getting closer to the surface, the tide going out, inching towards our flippers. Then we heard the boat sound its horn. Grace and I had been in the fish shoals for nearly two hours. We looked at each other, put our heads down and pretended not to hear the blast. I started to swim the other way and the horn blasted again. I saw Grace turn reluctantly and swim in. I kept going.

"One more minute," we used to scream to our parents beckoning from the beach as waves curved behind us, lifting our tiny bodies into the air before throwing us down, pushing us deep until the air left our chests and then letting us go – when we would wriggle and kick up to the surface, hair tossed like seaweed, bathers at times in a tangle around our ankles, and we would scream, again, "One more minute!" I heard the horn blast a third time and I knew this time it was just for me. One more minute, I thought, and dived deep, scanning, trying to swallow it all with my eyes: the coral, the chunky fish with knobbled brows, the crackling sound I imagined was the invisible machinery of crustaceans. There was the absence, too, for I'd felt it, despite my lack of expertise. There'd been turns, spooky corners where Grace and I ought to have come face-to-face with a creature, an unnerving marine intelligence, but there'd been nothing. I swam to the lip of the boat and took off my mask and snorkel, leaving it on the metal shelf and went under one more time. A little ritual.

The day I arrived in Townsville, a group of thirty people gathered on the wall of the Ross River Dam and prayed for rain. After three failed wet seasons, it was at less than 15 per cent capacity and the town was paying $30,000 per day to pump water from the Burdekin Dam. Long-running restrictions had seen many residents turn to groundwater to tend to their yards. The local bore guy was booked out for months. It was time to petition God for rain.

Gautam Adani has also seen his share of petitioning up this way. Against a glum backdrop of young adults standing beside main roads holding placards begging for work, this Indian billionaire, relatively unknown in Australia, is touted as the man who will turn things around. It seems that Townsville's recent falling out with another mining magnate, Clive Palmer, casts no shadow.

"Clive wouldn't be able to walk the streets here," one local told me, proudly and bitterly. Palmer, who gave away fifty-five Mercedes Benz cars and 700 overseas holidays to his employees at the Yabulu nickel refinery in 2010, is now a local pariah. The refinery that he took over from BHP eight years ago is in limbo, while Townsville City Council is pursuing Palmer for $700,000 in unpaid rates. Last year the federal government paid $73 million to Palmer's former employees. State bureaucrats, firefighters and police are now managing the risky refinery – where one smokestack is on the brink of collapse and its storage dams of toxic sludge are seeping into waterways.

But that was Palmer, this is Adani.

It was December 2016 when Gautam Adani flew into Townsville to meet the Queensland premier, Annastacia Palaszczuk, and the town's mayor, Jenny Hill. The front page of the Townsville Bulletin greeted him with a photoshopped image of his private jet coming in over Castle Hill, next to a "Welcome to Townsville" sign. Yes, there was some animosity; this is, after all, a democracy. A couple of hundred people gathered on the foreshore to protest against Adani's proposed coal mega-mine, and two native-title owners, Carol Prior and Ken Dodd, were also on his trail. So, a little friction. But for the nation's kingmakers, Adani may as well be Midas. That very morning he had nipped down to Melbourne to meet Prime Minister Malcolm Turnbull to discuss the Coalition's "conditional" offer to chip in $1 billion to his project. After the meeting in the Townsville City Council chambers, there was the obligatory handshake photo with the Queensland premier, and then he was gone – a "fly-in, fly-out" billionaire.

"You can't get the smile off my face," Annastacia Palaszczuk told the gathered reporters. Townsville was going to be the base for Adani's regional headquarters and remote operations centre. For the cameras, she added that she'd had a couple of wins that day. "The first," said Palaszczuk, "that [the] meeting was to be held in regional Queensland, in Townsville." Tick. "And secondly that we would be able to discuss local jobs." Tick. So many wins. Such leverage.

But there was more. Palaszczuk continued:

> I'm pleased to announce today that, following the meeting, I have got an iron-clad guarantee from Mr Adani that there will be no 457 visas as part of the workforce for this major project. I secondly have a guarantee of a Queensland First policy for jobs and especially for regional Queensland.

Not that it was in writing. The only assurance Queenslanders truly had was a photo of a handshake. The premier ought to have stopped there, but she kept going. The project, she said, would generate 10,000 jobs. "The life of this project will be anywhere between fifty and sixty years." In 2013 a television commercial for Adani Australia was broadcast across the state. A warm European woman's voice asked, "What does Adani mean to Queensland? A fresh stake in the global economy ... Adani means 10,000 jobs for Queensland workers [cue image of young man in hi-vis vest unspooling cables, smiling boyishly] ... Adani means $22 billion in royalties and taxes invested back into Queensland communities [cue hospital corridor and stretchers]."

Here's hoping Palaszczuk doesn't pin all her policies on the commercials she watches. Is it possible she is unaware of evidence given under oath by Adani's own handpicked expert, Jerome Fahrer, in the Queensland Land Court? A former Reserve Bank economist, Dr Fahrer works for consulting firm ACIL Allen, which was contracted by Adani to analyse the mine project's numbers. Fahrer was critical of Adani's previous consultants. The figure of 10,000 jobs, he told the court, was "extreme and unrealistic." At the peak of construction, Dr Fahrer estimated, there would be about 2400 workers. But there was a catch. Looking at the bigger picture, he found that these jobs would come at the expense of around 1500 jobs in manufacturing, agriculture and from other mines. This is important. For a premier under pressure to create jobs in a state with a population of 4.6 million, of whom more than 160,000 were unemployed in May this year, supporting a coalmine that will see job losses from elsewhere seems a serious folly. Overall, Fahrer said, the Adani mine and railway would create "on average around 1464 employee years of full-time equivalent direct and indirect jobs." In other words, a drop in the ocean. Supporters of the project are quick to point out that Fahrer did not include jobs associated with the port expansion at Abbot Point; however, Rod Campbell from the Australia Institute noted in the *Australian* that a large coal port expansion in Newcastle was expected to create no more than eighty jobs.

As for $22 billion in royalties and taxes invested in Queensland communities, the *Australian*'s Sarah Elks wrote that Adani estimates this will be generated over the proposed mine's first three decades – amounting to $730 million each year. However, the equation that led to this estimate is, of course, "commercial in confidence" – and the myriad companies that make up the Adani conglomerate make it nearly impossible to follow the money. What tax on what profits will be paid in Australia? How much will be siphoned off to Adani's "marketing hub" in Singapore and the Adani family company in the Cayman Islands? A similar mystery surrounds royalties, which is what the mining sector pays to the state for the stuff in the ground. The Minerals Council of Australia, a hefty lobbying group, refers to royalties as part of the "total tax burden." Considering that it is paid on the actual product, can it really be called a "burden"? Recent reports revealed the Queensland Labor government had split over an offer to Adani of a "royalty-free holiday" of up to $320 million in concessions – read free coal. In response, Adani announced it would postpone the decision to push ahead with the Carmichael project. In late May, Labor agreed to offer Adani a reduced royalty rate for the first few years, after which it will have to repay the deferred royalties plus interest.

With the tiniest tug, the promise of "10,000 jobs, $22 billion back into Queensland" is easily unravelled. Perhaps the simplest question is: will the Adani project put back into the economy what has been spent to support it? Will Queensland cover its spending, which amounts to more than $10 billion in the past decade alone, on building, maintaining and upgrading coal-related infrastructure? Is it even possible for a project to be both a source of massive revenue *and* a major recipient of taxpayer-funded infrastructure, deferred royalties and the like? So many zeros, so much alchemy, it could put you in a spin.

Palaszczuk continued with practised emphasis: "That means generational jobs. That means you can work on this project and your son or daughter may have the opportunity to also work on this project."

Um, "remote operations centre," anyone? It was as if the words, spoken less than ten minutes ago, had never been said. A few weeks earlier, the Australian CEO of Adani Mining, Jeyakumar Janakaraj, had confirmed all mining operations from pit to port would be fully automated. "In our minds, this is the mine of the future," he said – and it is. Rio Tinto and BHP have already started remotely operating extensive parts of their mines and rail connections. Journalists and academics Guy Pearse, Bob Burton and David McKnight revealed in their book *Big Coal* that in 2012 Rio Tinto replaced 380 train drivers with computers and Perth-based operators, proudly announcing at an investor seminar that robotics had so far saved the company from employing 900 people. *Guardian* journalist Oliver Balch earlier this year described the odd sensation of watching driverless trucks in Rio Tinto's Pilbara mines:

> Huge trucks trundle along the mines' reddish-brown terraced sides laden with high-grade iron ore. Back and forth, almost endlessly. Watch for long enough, however, and you'll see that no-one ever steps out of the cab. No lunch stops. No toilet breaks. No change of shift.

Today a BHP internet advertisement calls for the next generation to "Join our tech revolution." In it young adults of varying skin tones, some sporting a piercing or tattoo, appear happy and inspired as they perch on luminous white cubes in a space that looks very much like the Nicholas Building, a well-known hive of art studios in Melbourne. This is, as BHP says, the new generation of miners. Think university-educated "aspirationals." Think mining engineers, geologists, geophysicists. Think a bachelor's degree and at least five years' experience. Those kids in regional Queensland holding placards begging for jobs don't stand a chance. But then everyone knew that, didn't they?

*

Gautam Adani arrived on the Australian mining scene in 2010. His flagship company, Adani Enterprises, purchased Linc Energy's massive deposit of thermal coal in the Galilee Basin in western Queensland. Adani is one of India's most powerful men, and his company's vast influence is often linked to his proximity to Prime Minister Narendra Modi. Perhaps the most cited example of this is a photograph published in May 2014, on the day Modi was to be sworn in. Paranjoy Guha Thakurta of the *Citizen*, an Indian independent news site, explains:

> The photo depicted Modi leaving Ahmedabad, the biggest city in Gujarat, the state where he had been Chief Minister for nearly 12 years from October 2001 onwards, to travel to New Delhi to be sworn in

as Prime Minister of India. His stretched hand waving (presumably to an adulatory home crowd), the colourful logo of the private aircraft he was about to enter was clearly visible in the picture. It said: Adani.

It is no secret Adani and Modi are close. After Modi took office, it became a bizarre game of "Where's Wally?" as the new prime minister travelled the world inking international deals. "Everywhere that Prime Minister Narendra Modi goes, it seems, Gautam Adani is sure to go," wrote the *Hindustan Times*. But Modi did not make Adani: after all, one has to have millions to make billions.

The businessman's rise from sorting diamonds as a sixteen-year-old in his cousin's Mumbai business to earning his first million in the trade by the time he was twenty is the stuff of legend. From there, Adani went into the plastics business with his brother and by 1988 had started an import and export company, soon expanding into infrastructure and power generation. In 1998 Adani was allegedly kidnapped in his home city of Ahmedabad and held by underworld figures on instruction from a Dubai-based gangster. Of the experience, he told the *Australian*'s Michael McKenna, "They wanted a ransom and while we waited, I played cards with them. I am basically a calm person." The ensuing court case was shrouded in mystery; Adani had to be summonsed to give evidence against the accused. Today he owns a dizzying array of companies: interests in India include ports, property development, oil and gas exploration, coal mining and trading, coal-fired power stations, and cooking oil (think palm, soybean, cottonseed, nut plantations). Recently, Adani has also become India's largest solar power producer.

Now the proud owner of a thermal coal deposit in Australia, Adani plans to build a 388-kilometre railway stretching from Abbot Point port across Queensland to the Galilee Basin, under which lies one of the world's largest untapped coal seams. Adani plans to produce up to 60 million tonnes of coal from the Carmichael mine every year for sixty years. It would be the biggest coalmine in Australia. The coal is thermal, which is mostly used to generate electricity. The company's plan is to ship it through the Great Barrier Reef to its private port in India and feed it into its power stations. Waiting in the wings for Adani to start shifting "overburden" – mining-speak for bushland, gums and bottle trees, red dirt and aquifers, layers of roots and sediment that compressed the Permian-age plant matter into coal in the first place – are five more mining companies. The Galilee Basin has been divided into nine proposed mega-mines. Australia is currently the world's second-largest exporter of coal. At full production, the Galilee Basin is expected to double Australia's coal exports to more than 600 million tonnes a year.

Once Adani begins, the rest will follow – they are Chinese-owned MacMines Austasia, Brazilian giant Vale, Germany's Hans Mende's AMCI, Clive Palmer's Waratah Coal, and GVK Hancock, a joint project of Gina Rinehart and Indian billionaire G.V. Krishna Reddy. The coal seam would, when burnt, blow up to one-tenth of the world's total carbon budget – the amount scientists say we have left if we want to stop at 2°C of warming.[1]

In their way, holding the line, are a scattering of traditional owners and a grazier. Behind them is a vast network of willing activist bodies, conservation groups, academics and marine scientists.

The stage is set.

*

---

[1]   If the whole "degrees of warming" thing confuses you, you're not alone. A good way to picture it is this: think of your body as the planet and your clothing, heating and so forth as the atmosphere. Usually your body with the help of clothing, etc. will keep your temperature steady. This is the sweet spot humans have enjoyed on Earth for thousands of years. Now imagine you've got a fever. A fever of one degree, you're feeling flat, but you'll probably go to work. Two to three degrees, you are in bed at home. Take it to four to five degrees, you are not lucid and your body is losing water. Six degrees, you need to go to emergency, your organs might fail. It is an imperfect metaphor (is there such a thing as a perfect metaphor?), but I find it helpful.

"These cells multiplied in the oceans and some put out oxygen and this oxygen changed everything." I was confused. I'd asked the mayor of Townsville, Jenny Hill, if she accepted the science of global warming, making a conscious effort to say "accept" instead of "believe" – because in truth I struggle to *believe* in fax machines, let alone 3D printing, but I accept that these machines and their results are real, even if I don't understand the science and technology. Hill has a background in science, so I was aware the mayor could run rings around me if she wanted to, but I was trying to keep up.

"Are you talking about the beginning of life on Earth?" I asked in a squeaky voice. Hill nodded – a little sheepishly, I thought. Three and a half billion years ago seemed pretty far down the garden path to lead me. She quickly changed tack. "Technology," she surmised, "will overcome these problems."

There'd been a couple of these moments. Near the beginning of our interview, Hill had thrown in Aboriginal Australians for good measure. She had just visited Palm Island, she told me, where the locals were keen to work on Adani's mine in the Galilee Basin. "Indigenous people want to connect with their homeland," she said.

"By digging it up?" I asked – and honestly I did not say this to be a smart-arse. Australia's Indigenous people are long overdue a fair share of the nation's growing economy. But this argument, often thrown in the path of anti-coal activists, was all muddled up in Hill's office. I wasn't sure the desire to connect with one's traditional land was the same as mining it. As for the number of jobs Adani was going to provide in Townsville, Hill would not be drawn on an exact number. "I've a number I am comfortable with."

Then, after the Indigenous card and the Beginning of Life on Earth card, Hill finally got real. I asked about her moral responsibility to future generations of Townsville residents if the demise of the Great Barrier Reef is unchecked. In 2013 the federal government released a study on the Great Barrier Reef's contribution to Australia's economy. It found that the reef generated 69,000 jobs and $5.68 billion in annual turnover in 2011–12. Beyond this, cyclones, floods, fires, drought have always been a way of life in the north, but events in recent years have been relentless. Following back-to-back floods in 2009 and 2010 that claimed the lives of thirty-eight people, Queensland's then premier, Anna Bligh, noted, "Every time we settle into the task [of recovery], we get a new task added to our plate." The federal government poured $5.6 billion into reconstruction after these floods. What about those responsibilities? Hill, in response, looked me straight in the eye. "I've a community with a high rate of unemployment and a huge spike in youth crime. What about my moral responsibility to Townsville today?"

The town is certainly on an uneven keel. I'd walked the wide, hot and empty city streets. The image of a spaghetti western town came to mind as I passed big nightclubs, closed, some permanently, the odd shop doing limited trade – one closing down had a couple of backpackers picking over African trinkets. There were vacant and shuttered build-ings, teenagers' names written in the dust and grime, a huge chain chemist, inscrutable office blocks, a pub or two – tumbleweeds waiting for the next boom. A smattering of restaurants, most so sprawling that, even with customers, they looked empty. Occasionally, sliding doors opened as I passed, cool air gushing out, and I detected activity inside. Perhaps Townsville was like the desert, I thought, the wildlife bunkering down in the day and emerging at night. The crime statistics suggest as much. Last year a spike in "youth" crime saw an extra thirty police deployed at night. The local inspector had urged vigilante groups that were patrolling the streets to disband. A petition was circulated, demanding a curfew for teen-agers under sixteen. A cab driver told me about the constant car theft.

The wider Townsville was an unruly sprawl of suburbs. It took me days to understand the slow and lurching bus service that took between one and two hours to get from the city to James Cook University – one of the largest and fastest-growing institutions in the region, and recently hailed as the best university in the world at which to study marine and freshwater science.

Suburbs and malls had sprung up around the campus to service the 12,000-odd students. In Hill's office I suggested building a train line out to the campus. This, she said, was a decision for the state government. The council was planning to build a bus hub in the city.

Life, according to Hill, would be breathed back into Townsville through two projects. During the federal election last year, the council lobbied for a new rugby league stadium. In the 2013 election it had been all about West Sydney; this time round it was all about regional Queensland. Bill Shorten folded first, pledging $100 million. Turnbull followed suit, matching Labor's figure. On Twitter, the *Townsville Bulletin* wrote, "Regardless of who wins the Federal Election, Townsville will finally get its new stadium!" It could also be said that now we'll never really know if the people of regional Queensland wanted a stadium, let alone needed one.

"How will people get to the stadium?" I asked Hill, still stuck on public transport.

She looked at me as though I were crazy. *Same way they get to the stadium today, stupid,* I imagine she would have liked to have said. "They'll drive."

The second life-giving source was Adani. "We need Adani," Hill told me. "It's the next big thing."

"The only next big thing?" I asked.

"Yes," Hill said, emphatic.

Hill is pleased with the recent Adani announcement that Townsville will be the site of the mine's regional headquarters. It was hard-won. Last year saw a bun-fight among mayors up and down the coast vying for a piece of the action. The Whitsunday regional mayor had in October been given the green light at a council meeting to "immediately contact Adani chiefs and offer a parcel of land" in the township of Bowen, just south of Abbot Point.

Further south, Rockhampton's mayor offered to build Adani its very own airport to the tune of $22 million. Countering this, the mayor of Mackay proposed providing Adani with the town's own, already existing, airport terminal.

For its part, the *Daily Mercury*, the local newspaper for Mackay, which was purchased by News Corp in June along with seventy other regional newspapers in Queensland and northern New South Wales, ran a headline in December: "Put this in your diary if you want a piece of $21.7b Adani project." It was spruiking the Adani Roadshow, an event the mining company was putting on to drive interest in the project's construction phase.

Adani has sprinkled the prizes around, with local newspapers announcing Mackay will be the mine's "maintenance hub" and Bowen the "rail construction and operational services hub." But what kind of boom is it that you have to coerce into existence?

The previous state government had offered $450 million to help Adani build its railway. During the 2015 Queensland election, the Labor Opposition promised to protect the Great Barrier Reef and pledged to withdraw any state funding of Adani's railway. Queenslanders swung behind Labor and voted out Campbell Newman's Liberal Nationals. After six months in office, though, as Fairfax's Lisa Cox reported, the new state Labor government was considering alternatives to assist Adani, such as the "royalty holiday" mentioned above, in defiance of its own Treasury's advice that the Adani project was "unbankable" and inherently risky due to the company's opaque corporate structure and offshore entities – such as in the Cayman Islands. A year later, the state threw the mine another lifeline, this time giving the Adani project "critical infrastructure" status, which in turn grants the government the power to speed up assessments and block appeals. It is a status generally reserved for public infrastructure in areas such as health or national security.

How much more convincing does a boom need?

Well, in March this year, Premier Palaszczuk invited eight regional mayors to join her on a visit to India. The mayor of Gladstone, Matt Burnett, even overcame his fear of flying to go. You can see a photo of them in an ABC report all wearing blue hard hats inside Adani's Mundra Port. Three weeks later, Prime Minister Turnbull met Adani in New Delhi and assured the Indian businessman that native-title obstacles would be removed.

Now, sitting in the Townsville Council chambers, a huge modernist concrete building, Mayor Hill said, "We have offered to provide Adani with a temporary office while they looked at real estate for their head-quarters." It may have been the look on my face that did it, but Hill quickly added that Adani had declined the offer.

*

In New Delhi at the Global Natural Resources Conclave this year, Australian mining magnate Gina Rinehart stood onstage, clasping her book *From Red Tape to Red Carpet and Then Some* to her chest. The title, she explained, was borrowed from Prime Minister Modi, who once said, "No red tape, only red carpet, is my policy towards investors." She went on to congratulate India on its soaring success under Modi's guidance. Low inflation, vast foreign investment – "What an exciting country for entrepreneurs!" In Australia, however – Rinehart's tone became grave – it was a case of "the emperor fiddles while Rome burns." She wished it wasn't so.

Rinehart then spoke about all the permits and approvals she needed to get for her Roy Hill project in Australia. On a projector she screened a video of her friend Jim Viets, who played on his guitar a country and western song, "Mining Permit Blues." It's very good. So is red tape hold-ing back Australia? Or is it holding back Rinehart?

Three years ago, when iron ore prices began to slump, Rinehart issued an alert: "red tape, approvals and burdens" had to be cut. Jobs and revenue were in peril. Australia, it seemed, was in peril. Somehow she'd conflated her fate with Australia's. She was yet to send off her first shipment of iron ore from Roy Hill in the Pilbara, and the market had finally peaked and begun its descent. After a record stint of high prices, there was a glut, and bargaining power was significantly reduced. Roy Hill had missed the boat. Sure, it was still going to make money, but not what Rio Tinto, BHP and Fortescue Metals had been raking in. It was Australia's fault. All those permits. A lot of things are Australia's fault – red tape, taxes, the judicial system, employment entitlements, even the damage wrought by cyclones, as Queensland senator George Christensen wrote recently in the *Australian*. Queenslanders suffering the costs of Cyclone Debbie, he suggested, must lay the blame at the feet of Australians opposing Adani's coalmine. He listed the vast network of wealth and influence behind the anti-Adani campaign – millionaire bank CEOs, lobbying group GetUp!, anti-coal organisations with links to the Clintons, the former Labor minister Peter Garrett, the Wangan and Jagalingou Traditional Owners Council, those in their comfortable enclaves down south – in other words, the usual culture-baiting tirade most Australians have come to expect.

Either way, the argument is clear. Red tape is not merely an aspect of governance. It is a crisis point of two worldviews. Holding one worldview are those who see global warming as a profound threat and feel a responsibility to mitigate it. Within this is also a palpable sense of excitement about the opportunity the crisis may bring to re-set global priorities so that profit is not centrestage. It is this latter notion that those holding the other worldview fear. They do not feel threatened by climate change, but by the proposed solutions to it.

This is why so-called smackdowns on television programs such as the ABC's *Q&A*, the Punch and Judy show between the left and right, don't matter. It does not matter if British physicist Brian Cox, who is irrefutably better placed to judge climate science, tells One Nation senator Malcolm Roberts that global warming is real and human-caused. It does not matter because Roberts does not see the science; he sees a scenario in which all the Brian Coxes of the world are infinitely more powerful than all the Gina Rineharts. He sees a future in which a lauded Greens senator might say, *I told you so*. And this is scarier than a planet with a fever. When Rinehart says the sky will fall in on Australia if attempts are made to regulate and rein in the mining sector, she's right. Her idea of Australia – the country her father, grazier and iron ore magnate Lang Hancock, showed her, and what lies underneath it – is at risk. Laying out the red carpet isn't just about profit and growth, it is about entitlement.

Climate change lays all these bones bare. It cannot be summed up in a policy. It is a bright fluorescent light in the middle of the night from which no hypocrisy can hide. It is impossible to take a stand on climate change without making a value judgment – which is why politics has become so inert and why the Rineharts of this world are fighting with everything they've got.

Strategy number one – without mining, Australia is Nothing.

So just how important is the mining industry to our economy? In February this year, economics journalist Ross Gittins asked this question in the *Age*. "Short answer: not nearly as much as it wants us to believe, and has conned our politicians into believing." The next question is: why? Here Gittins took a hit for the team:

> Because people like me have spent so much time over the past decade and more banging on about the resources boom, we've probably left many people with an exaggerated impression of the sector's importance.

Couldn't have put it better myself.

Gittins went on to give the numbers: the mining sector accounts for 7 per cent of Australia's total production of goods and services (GDP) and employs 230,000 people, which is 2 per cent of Australia's workforce. Nothing to scoff at. If we imagined employment as a body, you could say mining is the hand. I like my hand. Yes, mining has a flow-on effect, but then again, so do I. Gittins pointed out that much of the profit and income generated by the mining sector goes offshore, which is fine, it's allowed – obviously. But that the mining sector is Australia's economic heart is less an argument than a pea-and-thimble trick.

Yes, there's tax, mostly at a rate of around 13 per cent, though Rio Tinto and BHP are at 27 per cent, but the Australian Tax Office claims a little bit of Singapore magic can get that down. There is also the question of subsidies – the Australia Institute says government subsidies to the fossil-fuel industry include $5.5 billion in non-agricultural fuel-tax credits, $650 million in tax deductions for exploration and prospecting, and $1.24 billion in concessions on aviation fuel each year. In response, mining lobby groups have hotly taken issue with the word "subsidy" and argued that fuel credits are exemptions from "road user" costs which are granted because mining companies build and maintain their own roads. As for the exploration and prospecting deductions, let's just say this is more in the red carpet than the red tape category.

Queensland is Australia's largest coal exporter. It has five working coal regions: the Bowen Basin in central Queensland and the Surat, Clarence-Moreton, Tarong and Callide basins in southeast Queensland. In 2012, when a United Nations report urged the Queensland government to step up its protection of the Great Barrier Reef, particularly from port development, Premier Campbell Newman was unmoved. "We are in the coal business. If you want decent hospitals, schools and police on the beat, we all need to understand that." A year later, in 2013, Queensland's Treasury refuted the claim that the mining wealth is shared around the state. In a report, Treasury stated, "Governments face budget constraints and spending on mining-related infrastructure means less infrastructure spending in other areas, including social infrastructure such as hospitals and schools."

But Newman was on the money about being in the coal business. There are fifty coalmines currently operating in Queensland and a further twenty-one, including in the Galilee Basin, in the pipeline. As well as the billions spent on coal-related infrastructure, there have been concessions made on water and electricity supplies and substandard bonds for mine rehabilitation. The government has also waved through three LNG facilities on Curtis Island, in the Great Barrier Reef near Gladstone, and approved a rapid, at times shonky, rollout of coal-seam gas wells (royalty per well) across the state, and unconventional gas projects, which ignite underground coal seams to generate gas, with sometimes tragic results.

The revenue must be huge. But it isn't. Sometimes it's a good innings and sometimes it's a very bad innings. Last year royalties came in at $1.59 billion, 3 per cent of the state's revenue. To compare, car registrations came in at $1.63 billion. But the state is expecting good things this year; politicians are hoping for around $3 billion. It still doesn't make sense – but it all adds up, and it has to *keep* adding up. Spending has been locked in, promises made, factions formed, donations offered – it's hard to give up.

And so, the voodoo continues.

"This is," thundered Prime Minister Tony Abbott in 2015, "a project that will create 10,000 jobs." Yes, after Fahrer's evidence. He may as well have dug a hole and whispered his evidence into the ground. *It is sabotage!* bellowed Abbott of the court cases piling up against Galilee Basin projects – there have been farmers concerned about water, something about a bloody skink and a snake, a fool trying to link coal and climate change, and damn native title. For coal's sake, Brandis, change the legislation.

Andrew Morrell is a big guy. I felt like a sparrow flitting beside him as we walked along the corridor in Bowen TAFE to a meeting room. I'd driven through countless cane fields to get there, past billboards for fertiliser saying "More Cane, More Sugar, More Profit," and through the Caley Valley wetlands, which regularly teem with birdlife from all over the world and are an important filter for four of the five Great Barrier Reef catchments. A quiet town of 10,000, Bowen – like so many towns up this way – once heralded itself as the "capital" of North Queensland. I peered into people's gardens and saw row upon row of taro plants. Surrounded by fertile soil and farms, it is a 457-visa town. Pacific islanders make up the bulk of crop pickers. Just north of Bowen, next to the wetlands, is Abbot Point port, with its six huge stockpiles of coal. In April this year the wetlands turned black from coal spillage after Cyclone Debbie went through the region. On the other side of the port is the reef.

Shadowing each stockpile is a stacker-reclaimer, 30 metres tall with huge mechanical arms, that shifts coal from wagons and pours it into mounds. The trains are over two kilometres long, shuttling back and forth from the Bowen Basin, a vast inland reserve of mostly coking coal – used with iron ore to produce steel, with nickel to produce stainless steel – and some thermal coal. The Bowen Basin contains about 70 per cent of Queensland's coal deposits. From the stockpiles, the stacker-reclaimers scoop up the coal again and pour it onto a three-kilometre-long conveyor belt that heads out to sea, where the ships wait.

When locals claimed in February that coal was washing up on beaches near the port, the story quickly did the media rounds and came up on my phone during an interview with a former director of the Great Barrier Reef Marine Park Authority. He laughed with a kind of bittersweet delight when I mentioned it. "We've known about that coal spillage forever. The ports would get fined every year. It goes out on a big conveyor belt, falls off and washes up. Cheaper to pay the fine." Carol Prior, a Juru traditional owner whose country abuts on Abbot Point port, said the same thing. "Oh yeah, you've always been able to see it. You can even see the dust sitting on top of the water."

Andrew Morrell is also a Juru man. His hands were like huge paws. They opened and closed as he talked, palms rolling over, cupping air. He taught motor mechanics, and we talked cars for a while. Reminisced about old Holdens. Morrell told me the course he was teaching was set to shift online. He couldn't imagine not being able to talk to his students directly. That is, if the course survived the next round of reforms and funding cuts. TAFE, it seems, is always under threat, though amid the constant claims of skill shortages and high youth unemployment you'd think supporting apprenticeships would be a no-brainer. Morrell talked about trying to teach the students to recycle old parts, how to make their own tools. "Like we used to," he said. "These days you take a part off, throw it away and put a new one on." I asked Morrell if he could take me out to part of the Juru country, near the port, and he sighed. Shook his head. Checked his watch. "You can go, if you like," he said kindly. "You're allowed." Morrell explained he and his family used to go there and walk along the middens in the dunes – discarded shells and sharpened rocks, the remains of his people's long-ago feasts. Morrell's father would stop to collect bush tucker and then go onto the beach where he would fish as a child. Further along is a Juru burial ground, with graves dating back thousands of years. But now they were always stopped by security. "They file a report to the police. Nothing happens, but it's not a nice day out."

In 2011 Adani paid $1.8 billion to the Queensland government to lease the terminal at Abbot Point port for ninety-nine years. There are plans to expand the port so that Adani can increase its holding capacity, while GVK Hancock is proposing to build another terminal. All of this will see a road and another six stockpiles, each 50 metres wide and running for a couple of kilometres, carved into Juru country. The first length of Adani's proposed 388-kilometre railway to the Carmichael mine will also be laid down here. The middens and rolling dunes, which served as a meeting place for many clans, are all set to go. The Juru people, including the Morrell family, were ready to accept this. They entered into negotiations with Adani, open for business. Which is lucky, because under the *Native Title Act*, there is no "right not to negotiate," only the "right to negotiate," albeit for a limited time. In 2011 an ABC *Four Corners* investigation into such negotiations spoke to a native-title expert, Ciaran O'Faircheallaigh, who had attended hundreds of meetings between mining companies and Indigenous groups. O'Faircheallaigh described the inequitable power dynamic:

> When a mining company sits down to negotiate with native-title parties there is a six-month period of negotiation. The mining company knows that at the end of the six months it can go to the National Native Title Tribunal and get its mining lease. This is a very simple and fundamental point. If one bargaining party is under enormous pressure to do a deal and the other one isn't, the people who are under pressure generally have to give in, and that's what's happened and that's why, except where Aboriginal people have major political power, deals tend to be very uneven.

For the Juru people, the negotiations fell apart almost as soon as they began. But as often happens when such negotiations disintegrate, it is not the company that fractures – it is the traditional owners. There is division, as some want to take a deal, no matter how paltry, because once a company takes them to the Native Title Tribunal, they may get, quite literally, nothing. By law the tribunal can only grant interests – that is, mining leases, which it approves 99 per cent of the time – and is not allowed to consider payments for the minerals extracted from native-title land. Given this, some claimants might want to hold off, to negotiate harder, to get what they see as a fair deal.

A company might stand back here and let the divided parties sort themselves out, or it might start to get in someone's ear.

It is, as many say, an age-old strategy – divide and conquer. Money starts to flow – the Act rules that companies seeking an agreement must fund the negotiations, which seems fair until you start wondering where the line is drawn. What if they fund some members and not others? There are authorisation meetings to decide which members will represent the wider group and there are unauthorised authorisation meetings, there are representatives who are more representative than other representatives, new claimants are added to claims, old claimants struck off claims, there are breakaway factions, sometimes flush with funds, there are bus-loads of people brought in for votes and differing views on just who they are – family or ring-ins or both. And hey presto, out pops an agreement.

It didn't happen like that, say some Juru people of the Adani agreement. It did, say others. Juru member Angelina Akee, chair of Kyburra Munda Yalga Aboriginal Corporation, which authorised the agreement, says the Juru are looking forward to helping Adani achieve its 7.5 per cent Indigenous jobs target. Others scoff. The group is bitterly divided. Members say they have been shut out and accuse Kyburra of refusing certain memberships, failing to keep proper minutes, not providing accurate financial records or a true record of payments received, including more than $2 million from Adani. In August last year, Juru members lodged a complaint with the Indigenous corporations watchdog, which is now conducting an investigation. This follows a separate review of Kyburra by the Australian Charities and Not-for-Profits Commission over its management of an Indigenous cultural centre in Townsville. It is alleged Angelina Akee authorised three questionable payments into personal bank accounts, including $4000 to her daughter, Tanya Akee. Angelina Akee's sister, Agnes Tapim, was on the cultural centre's board; her husband, Frances Tapim, on the Trust board. Juru members have held protests outside the centre and accused those in charge of nepotism. In May this year the ACNC decided to revoke the cultural centre's charity status.

Andrew Morrell looked at me sadly. He said his family and other members were either ignored or shut out of the cultural heritage planning process, which Adani had transferred $825,000 to Kyburra to undertake. He and Carol Prior asked for a 500-metre buffer between the port expansion and the burial ground. It is five metres. But while Carol has been forthright in talking about this to reporters, Morrell said he was nervous. His father had been a crane driver putting up windmills when Premier Joh Bjelke-Petersen told Queenslanders after the *Mabo* ruling that the Aborigines were "coming for your backyards." His father often told him about the time he watched a bunch of "cattle cockies" bulldoze one of his family's sacred sites.

Again Morrell's hands rolled into each other. At least he knows his story, he said to me, even if there are gaps – like he knows where places are on country, but for some he doesn't know their meaning. "But there are people out there who don't even know their country."

*

"Indigenous Australia is open for business," the formidable Marcia Langton wrote in the *Guardian* last year. She listed various ways Indigenous Australians have fought their way into the Australian economy, noting the existence of 3000 Indigenous businesses, timber mills, carbon-trading and cattle stations and about 2000 agreements, resulting in "many strong part-nerships between private developers and traditional owners." Langton also urged traditional owners resisting economic development to take their feet "off the brake." Though Langton's call for economic determination was nuanced and complex, it didn't taken long for it to be kneaded and rolled into the overarching narrative Australians have been bombarded with since they voted for action on climate change in 2007: mining is the back-bone of Australia, and without it we're doomed. It was an easy enough sell – especially after the global financial crisis – and it saw many urban Australians looking at hospitals, schools and train stations and wondering if the mining boom had paid for them (answer – no, it didn't. We are all in our many guises the backbone of Australia). You would think, however, that this would be a more difficult case to make in remote communities where there was practically no infrastructure. It would hardly be good PR to say that mining paid for a dilapidated health clinic and a shipping container for a school, when in fact mining, while often occurring on native-title land, had completely passed many Indigenous groups by.

In the *Monthly*, Cape York leader Noel Pearson said as much in 2015:

> Think about it. Just before Australia entered into one of the greatest mining booms in history, which lasted the best part of two decades, the High Court's 1992 Mabo decision recognised the land rights of Aboriginal people. The legislative protection of native title under Prime Minister Paul Keating's *Native Title Act* in 1993 and its subsequent expansion three years later in the Wik decision should have meant that traditional owners across the continent were now in a prime position to partake in the economic development of Australia. Never before in the history of Australia were its indigenous peoples better placed.
>
> It didn't happen.

Rubbing salt into the wound is the oft-stated claim that Australians avoided the aftermath of the global financial crisis. In the six months after the collapse of Lehman Brothers in New York, Australia's mining sector shed nearly one-fifth of its workers. In resource-rich Queensland and Western Australia, the Indigenous unemployment rate almost doubled. There are two views here – that this is a case of "last on the bus, first off the bus," or that this shows the importance of mining jobs to Indigenous Australia. Langton estimated there were 7000 Indigenous workers in the mining sector in 2012 – a significant number, especially in remote communities – although overall this represents only 4 per cent of Indigenous employment. Pearson again:

> Many companies have been constructive in supporting this emergence. But the aggregate story is one of an enormous swindle perpetrated against Aboriginal Australians.

There is more. This enormous swindle includes not only mining companies, but also conservation groups, in what Marcia Langton calls "a strange twist on the racist fiction of terra nullius." In her scathing 2012

Boyer Lectures, *The Quiet Revolution: Indigenous People and the Resources Boom*, Langton said in her steely way: "They are not wilderness areas. They are Aboriginal homelands, shaped over millennia by Aboriginal people."

At first it appeared there had been a radical shift in alliances out on country – and the bolt of fury levelled at the environment movement caught left-wing ideologues off-guard. They had assumed a "natural" alliance with Indigenous Australia. As sleek as mercury, mining companies quickly leveraged the new fury and began to forge the impression that Indigenous Australia now shared its quarry vision. This was helped along by Langton's public endorsement of Andrew "Twiggy" Forrest of Fortescue Metals. Yes, the line went, many Indigenous communities may have nothing to show for the boom – but nor do they have anything to show for a failed era of welfare. Indigenous Australians now understood that quarry vision was their only way out of the wretched and stricken hole they'd been dropped in by the left. It is an old move out of an old playbook – and it has hijacked real jobs and pride. It's "Let's make black lives political pawns in the eternal battle between left and right" so that the usual suspects can turn the battle in their favour.

But this is not as simple as a backlash against the left or the right.

When Palaszczuk came to power in Queensland, she called a halt to land clearing across the state and felt the full force of Noel Pearson's awesome wrath. For two years, pastoralists and farmers had dragged thick chains between dozers across their land to remove what remained of the natural vegetation. The previous premier, Campbell Newman, had controversially loosened regulations, triggering a rush of voracious plucking that saw Queensland reclaim its title of Best Land Clearer in the nation. Now the new state Labor government wanted to rein it in – an ultimately futile and predictable gesture to appease the left while simultaneously waving through one of the world's biggest coalmines. Pearson mounted a fierce challenge. Lefties were appalled. Oh no, not him too, was the sinking feeling: a dread that Pearson and the Cape York mobs had joined the ranks of climate change denial.

He hadn't. They hadn't. Pearson just wanted the government unequivocally and quite legitimately to fuck off. His fury is a hot growling rage from a people backed into a corner and then held there. Again in the *Monthly*, he wrote that Cape York retains more than 98 per cent of its native vegetation. "It is probably the region's largest economic resource, its greatest value could possibly lie in its long-term preservation, but this value is being destroyed."

Pearson is not saying the people of Cape York necessarily want to clear it. But they'll be damned if they cannot benefit from keeping it. A state law to "protect the land" would not only annul native-title rights but also shut traditional owners out of yet another, more modern economy: carbon trading.

He continued:

> Contrast the traditional owner in Cape York with the white pastoralist in Central Queensland. The lands in Cape York have hardly been cleared. Meanwhile, the pastoral properties in the mulga country were cleared by ball and chain sometime over the previous century. The pastoralist enjoyed the returns from the old, dirty economy of the past. Those who had no foothold in the past have no foothold in the future.

For decades there have been calls for reform, to make the *Native Title Act* more equitable, and for taxation reforms so that traditional owners can share an economic inheritance most other Australians take for granted. They have all gone unheeded.

*

At the other end of the rail line to Abbot Point, 388 kilometres in, is Wangan and Jagalingou (W&J) country. In 2004, with a history that stretches back thousands of years, a few families put in a native-title claim to the area. Their connection report was extensive, detailing bloodlines and sites, seeking a return of their country.

To date, they have not received a determination. Or a dismissal. Native-title claims are known to take a long time, even up to a decade, but I imagine this particular claim is a hot potato, with lawyers pushing it around, hovering their mouse over it, then past it. Who would want to touch this claim? For even though it looked like nothing was happening on that land in 2004 – but for bottle trees, spiky echidnas and the odd rustle of cattle – plans were afoot. In 2009 Linc Energy announced on the ASX that it had drilled four holes, 120 metres deep. There was, on Wangan and Jagalingou land, a shitload of coal.

In 2012 along came Adani, and depending on how you view these things, the W&J people were going to get rich quick or lose their land, again, before they had even got it back. Again there was the "right to negotiate" and the W&J people squared up. The wider claim group put forward family representatives to negotiate, and in a unanimous vote they decided against the mine, which would turn their country into an "open void" (mining-speak, not my own), a 280-square-kilometre coalmine. In their unresolved title claim, they had applied for the right to camp, fish and be buried on country – there was no case here for coexistence. Adani did not blink and lodged an action with the tribunal. As expected, Adani was given the green light, granting it two mining leases. The mine, the tribunal declared, was in the public interest.

But Adani still had one more hurdle, one the W&J had been betting on. Adani needed a third lease from the W&J people, this time for the land on which they were planning to build an airport, workers' camp and power station. This lease was trickier because Adani needed parts of it to be converted into freehold, meaning the W&J people had to agree to surrender their native title forever. As for the "open void," they'd be allowed to keep that when Adani was finished. If the W&J people continued to refuse to negotiate, the mining company could go to the tribunal for the lease, but to get the freehold they would have to rely on the state government to compulsorily acquire the land. Which would cost the state a lot of money – and would look bad, whereas an agreement is "clean," something to promote.

Adani kept up pressure to gain an agreement. In 2014, more W&J people had added their names and histories to the native-title claim. There were now twelve families in total. They came together to vote on granting Adani the third mining lease and surrendering their land. The vote came in 93 against, 78 for and 4 abstaining.

From here, things got predictably complicated. Adani went to the tribunal for the third mining lease and continued to manoeuvre around the "roadblocks," as their lawyers put it. Soon there were two groups: the original claim group, the Wangan and Jagalingou Traditional Owners Council and Cato Galilee, a company with two directors. Both, of course, say they represent the will of the W&J people. Now, if you read the *Australian*, you'll get the impression that the Council is the breakaway group. If you read the *Guardian*, you'll get the opposite impression.

In 2015 another meeting was organised. At the time of their claim in 2004, the W&J people had voted for two groups – a family representative group who picked an applicant group, which was then formally authorised by the people. The applicants were to advance the native-title claim through the legal process and manage negotiations, and the family representative group was to ensure the applicants were acting fairly and in the wider interests of the W&J people. However, by 2015, the majority of the family representative group (18 people representing 9 of the 12 families) claimed to be unhappy with the applicant group – in particular, with two individuals who were refusing to accept the majority vote. This meeting aimed to break the deadlock by voting for a new applicant group. But members of the W&J Council were surprised to see Adani representatives when they arrived. As Adrian Burragubba later recounted to journalist Michael West, "We saw the buses turn up and we were wondering what was going on." The W&J Council said Adani had organised three days of accommodation, food and travel for 150 people connected to the two directors in Cato Galilee. The Council say they were unprepared. They had been expecting a smaller meeting and did not have the resources or support to organise for all the family members associated with the Council to be present, many of whom were dispersed through-out Queensland and New South Wales. And now Adani announced a vote on a memorandum of understanding they had prepared with the directors of Cato Galilee.

The W&J Council representatives decided to fight. Alongside the Cato Galilee directors, they made their case to the larger group. Amazingly, they won. The idea of a MOU was rejected and Adani's representatives were asked to leave. But it was working, the margin between votes was getting smaller, and immense damage was done. For those drifting between the two groups, there was confusion about meetings and communications – which meetings were authorised, and which weren't? And again, money was flowing.

The W&J Council started crowd-funding and lawyered up, filing four cases in the federal court. Among their members are Tony McAvoy, Australia's first Indigenous senior counsel, Adrian Burragubba, a musician, and 22-year-old Murrawah Johnson, whose family opened the Bowman Johnson Hostel in South Brisbane in 1992, which provided emergency accommodation for Indigenous youth and adults. The W&J Council also told native-title lawyers to remind the Cato Galilee company of the rule they'd agreed on: that all monies would be paid into a trust for the W&J people. In the Guardian, Joshua Robertson reported that the two directors of Cato Galilee flagged keeping the money as compensation for lost wages and use of their "intellectual property."

In 2015 Murrawah Johnson embarked on an eighteen-day world tour with her uncle Burragubba to meet with banks such as Goldman Sachs, Citibank, Bank of America, Credit Suisse and Standard Chartered. Outside Standard Chartered in London, Burragubba told a handheld camera, "We've come a long way to talk to the people in this bank. A lot of people don't know about Aboriginal people." Burragubba shrugged wryly. "Well, it's a bank." The London bank had invested $680 million in Adani. "They refused our request to meet with them," Johnson later told 4ZZZ FM, "and we were, well, we're coming anyway." Finally, a meeting was granted and it was, Johnson described, "welcoming." The two traditional owners told the bank that the W&J had not consented to the Adani mine and the bank had an obligation, under the UN Declaration on the Rights of Indigenous People, to take that into account.

"We have something to offer, we are not destitute people, we're not people without any belief," Adrian Burragubba said afterwards in the dull grey street, again to the handheld camera. "But if a mining company like Adani goes in there and rips the heart out of our country, there is nothing left and we'll have nothing to share with the world."

A month later, Standard Chartered cancelled its contract with Adani.

A groundswell of support was gathering behind the W&J Council and the families they represented. As for Cato Galilee, Adani and the state government appeared to have its back. The ABC obtained documents from the Queensland coordinator-general's office showing the state was intending to forcibly acquire the land. But it didn't need to, because in April last year, Adani announced its relationship with the W&J people had turned a corner.

There'd been a vote, 294 to 1, in favour of the Adani project.

The W&J Council put out a statement saying it was a sham, the vote was a stitch-up – and you can see why: the numbers didn't make sense. The W&J Council represented nine families – and yet only one voter stood against the tide? The implication was that it was W&J Council spokesman Adrian Burragubba, apparently just one disruptive guy wrecking the party. Warren Mundine, a former politician and Aboriginal leader, wrote in the *Australian*:

> One individual, holding himself out as representing the group, challenged the ILUA [Indigenous Land Use Agreement], demanding unanimous approval.

Contrary to this, the W&J Council stated that they and their families were not at the meeting: the vote – if it had even happened – had gone ahead without them.

"It's been reported," Mundine continued, that "the challenge has been financially supported by an activist group, partly funded out of the US."

In October last year WikiLeaks released a series of emails. Among them were revelations that the US-based Sandler Foundation funded the Sunrise Project, which had in 2014 offered the W&J Council financial assistance and scholarships.

The outrage was palpable. Michael McKenna wrote in the *Australian* that Cato Galilee's Irene Simpson had "long-held suspicions of the foreign funding of activists who had infiltrated her clan." But Simpson also noted that the Sunrise Project's offer wasn't so secret after all – the group had come to a meeting and spoken about why opposing the mine was important. "Then there was this with the Sunrise Project, which we always thought had some sort of overseas funding." What, like the Adani project?

The W&J Council stated they had not taken up the Sunrise Project's offer.

Cato Galilee put out a statement saying it was looking forward to working with Adani. The agreement, which is waiting to be processed by the tribunal and is now the subject of litigation in the federal court, has fifty jobs for the Wangan and Jagalingou people in a proposed bus company at the mine. "I don't want to drive a bus," said Murrawah Johnson to reporters. "I don't want to drive a bunch of FIFO guys to the biggest hole in the world." As for the price of the land surrendered, $398,750, the W&J Council reflected that the fauna got a better deal.

> Adani ... will compensate for the loss of 9,700 hectares of habitat for the Black Throated Finch by creating offset areas of 30,000 hectares, yet they do not offer one scrap of land to us for the extinguishment of our native title.

Two months later, the federal environment and energy minister, Josh Frydenberg, cornered US energy secretary Ernest Moniz to talk about the importance of the proposed Adani coalmine and the interference of American charities offering to support local groups that opposed the project. They were at the climate change talks in Morocco.

*

The W&J Council say they want sovereignty, true land rights. In his article in the *Australian*, Mundine likened the Indigenous Land Use Agreements to treaties, arguing there was "little difference." This is a stretch. Even for those with political clout, as the Wik and the Wik Waya people in Aurukun, Cape York learnt, there are limits.

In the early 2000s, leaders in Aurukun decided in a flash of brilliance that they did not want to negotiate with mining companies – they wanted to be the mining company. The Aurukun Bauxite Development (ABD) was formed; native-title holders retained 15 per cent equity and had two seats on the board, and would have 70 per cent of the 500 jobs. Nick Stump, formerly CEO of Comalco and Chief Executive of MIM Holdings, was the chair. ABD demonstrates how a mining company ought to do business on native-title land. Its proposed mine was planned from beginning to end, including rehabilitation, and information was translated into Wik-Mungkan, a local language. In 2014, the Queensland government rejected tenders from ABD and scandal-addled Glencore to start mining in Aurukun, but then secretly opened the tender for one day only for Glencore to reapply. The Wik people only learnt of the tender and subsequent licence six months later through Glencore lawyers. Since then the new Labor government, while seeming to concede a whiff of something rotten, has refused to overturn the decision. And in Aurukun? Tensions are spilling over, as the Wik and Wik Waya people find themselves with the "right to negotiate" with Glencore.

As for Adani, the company announced it had signed agreements with all the traditional owners from port to mine, and the state threw in "critical infrastructure" status. It was all slotting into place. In a couple of the news reports, there was a tiny bit of room left at the bottom for a single Indigenous voice.

"Shame on you," said Carol Prior. "Shame on you."

I turned the car inland, ocean at my back, and drove into the Bowen Basin. This was coal country and everyone was here, among them Anglo American, BHP, Rio Tinto, Peabody Energy, Glencore. Underground, the coal seam starts in Collinsville and stretches southwards for 600 kilometres, tapering off beneath Theodore. The Galilee Basin seam is further west, beyond the Bowen Basin, but as it extends south, the two seams meet. Looking at a map, you see them spill, bulge, split and curve towards each other. It is like rain on a window. Out there in the Galilee, the new kids on the block had pegged out their deposits – Palmer and Rinehart, Indians, Chinese.

It was humid; the sky was low. On the radio, farmers called in to report rain and you could imagine the sighs of those listening. There were farms in the region too, some with open-cut mines just a short buffer from their crops. I saw a mechanical grinding centipede of coal on the tracks beside the road and peered out, trying to see if there was a driver. See if it was a ghost train. A man was speaking on the radio now, from Hobart, railing against abortions. I stabbed the radio off. Eventually I drove through Collinsville, "Pit Pony Capital," the highway winding into wide empty streets, chunks of coal fastened to the ground, and murals of miners and horses coming out of pits. But no one had a map to buy. Nothing I could unfold and trace my finger along to get the lay of the land. I had to rely on my phone, the robotic voice issuing orders from my lap. It started to rain and I pulled over, stood in it for a while. I was trying to shake a feeling. A dozen or so roads kept T-boning with the road I was on: access roads with big metal gates, leading to some of the Bowen Basin's fifty or so coalmines. Every now and then you got a glimpse, a metal mouth poking into the sky, enormous pits like a hollowed-out marble cake. Scraped-out, endless voids. Maybe it wasn't such a bad word for them after all.

I obeyed my phone until I'd driven for an hour. "Turn left." A sign greeted me: "Private Road." I did that thing with my fingers to zoom out on the map and it enlarged too quickly. Then it shrank too quickly. I hated it. I didn't want to drive back. Another sign told me I wasn't allowed to be here. I pretended I didn't see it. Half an hour in, the bitumen stopped and the road turned to red dirt. The sides of the road fell away. An open-pit mine was being extended. In front of me were trucks and graders. I stopped the tiny white hire car and a man wearing orange hi-vis clothes walked to my open window.

"I have no idea where I am," I said.

He laughed and pointed to three yellow machines. "See that grader? Go around it, then past the trucks."

Right, the grader. I took a punt, drove up the slope, tyres spinning a little, then down. Another hour and I drove into Moranbah. Red Greyhound buses dropped men off and picked them up. There were motels, and a slab of dongas in the middle of town – tiny white windowless boxes, the only sign of life a pair of tan steel-cap boots at the door. In my motel room, I opened the curtains. There was a tyre yard, a couple of guys rolling the huge rubber beasts around. I wedged open the window. The tyres stank. I closed it. I opened the door again. Left it open and lay on the bed. I was still trying to shrug off the feeling of the road, how it was more like a bitumen bridge over percolated land. It was incredible in its way, the efficiency of all those vast gaping holes behind the buffer of straggly trees. The Institute of Public Affairs' Sinclair Davidson said it was "mining snobbery" when the Labor government in 2011 began spruiking its "education revolution":

> By linking the end of the mining boom with the need for an education revolution, the government is somehow implying that mining is an unskilled or semi-skilled activity. Similarly, by claiming that the end of the mining boom creates a need for more research and development the government implies that mining is somehow low-tech.

He had a point. There is a sense of concentration out here, a grim efficiency and understanding of the mechanics, not just of machinery, but of the seam itself. I'd driven for over 300 kilometres and never had the thought that the operations were semi- or unskilled. But it was ghostly. There was no life, no people, no quirky roadside stalls, no line-up of mail-boxes made out of old milk barrels, no farmer's trust tin. Not even roadkill. Nothing.

*

Moranbah ebbs and flows. It is a mining town, after all. Across the Bowen Basin, in a span of ten years, locals saw the population more than double from nearly 23,000. Old-timers had seen it before – there'd be a flood of workers for the building phase, and then they'd be gone. Newbies, however, were pumped. House prices jumped from $150,000 to $750,000, while weekly rents for an unremarkable weatherboard house skyrocketed to $1000 as mining companies fought to put roofs over workers' heads. Developers knocked down houses and put up units. Across Peak Downs Highway, streets and lights were put in, then new homes. Motels rose up out of the ground.

Property investors came from all over. Consider the Moloneys, who, as Trent Dalton wrote in the *Weekend Australian* in 2016, snapped up "their slice of the Moranbah boom":

> Kate and Matt Moloney, the savvy 20-something couple from rural Victoria, [were] crowned 2012's "Investors of the Year" by *Your Investment Property* magazine after turning a string of 13 Moranbah house purchases into an $8.5 million property empire. "These are ordinary, everyday Australians who have chosen to make a difference in their lives through property investing," gushed the magazine.

Three years on, they're bankrupt. The Moranbah Community Workers' Club, a bar and bistro, got a $5 million renovation. It's huge and mostly empty. The mines' building phase was finished. The houses emptied and the digging began. Reports began to pop up on the news that Moranbah was "hurting" – and for a while this was attributed to the end of the boom, but it became clear that beneath the decade's activity Moranbah had been hurting all along. The modern mineworker is a ghost. Fly-in, fly-out (FIFO) – or drive-in, drive-out – is a practice that began, legitimately, in remote areas of Western Australia, but has seen a number of towns in New South Wales and central Queensland warp in unusual ways, earning FIFO the nickname "cancer of the bush." Moranbah is more a camp than a community these days. So much activity and yet shops close down, the local footy team struggles for numbers, school fetes are a subdued affair. Mines have started switching to compulsory FIFO, which means if you live locally and want a job, you have to fly out to, say, Brisbane, and fly in with everyone else. Even though home, your family, is just there.

Across the Bowen Basin, there have been rollouts of thousands of dongas, creating self-sustaining universes with caterers and canteens, where workers do twelve-hour shifts, sometimes hot-bedding it, nine days on and four days off or a variation on that – and part of the days off are spent driving or flying. The mine is on twenty-four hours, seven days a week. Which is lot of time but not enough apparently to put covers on the hundreds of wagons travelling to and from the ports, to minimise release of coal dust. In the last two years, nineteen miners in central Queensland have been diagnosed with black lung, a disease caused by exposure to coal dust – and studies keep coming out of the upper Hunter Valley showing abnormal rates of poor respiratory capacity in children. There are also the extremely high levels of lead in generations of children in Mount Isa and Port Pirie. And yet for many communities these mines have become indispensable. Quietly, other jobs there have been edged out.

But since the mining downturn, thousands of jobs have returned to Queensland in agriculture, health, tourism and education, according to the Australian Bureau of Statistics. Contrary to all the alarmism, the economy is in the process of recovering and growing. Lucrative as it was, the coal rush wreaked havoc on other industries, the high dollar pushed manufacturers to the wall, and high wages saw a rush of workers redistributed to the Bowen.

But here, the uneven keel continues. Farms have been bought out and the mines, ever-expanding and gnawing, are starting to outgrow their towns. So much so that some may soon feel the mining companies' gaze on them. Here's hoping Moranbah doesn't have high-grade Permian coal beneath it.

*

"Well, it's a long way from Hollywood to Hay Point, but that's where I am right now. South of Mackay, Queensland." It's 1975 and actor Rod Taylor is standing atop a coal conveyor, his name stencilled on a hard hat. Wagging his finger, he reminds viewers, "The Utah Development Company built Hay Point." Cut to another commercial and now Taylor is on the streets of Moranbah. He cocks his head as if a thought just struck him: "It takes a lot of people and a lot of jobs to keep these towns running and that's not just up here. Think of all the other people in other jobs all over Australia needed to keep these people living." He wags his finger again: "They tell me, for every one job created by Utah, three others are created in other industries. A lot of people benefit from Utah backing Australia."

The Utah Development Company built Moranbah – it was one of their oft-cited achievements. Back then, mining leases tended to be conditional on the companies building or financing local community infrastructure. But even then, not long after the town's establishment in 1969, unease began to grow. Just how much was Australia getting out of this deal, people began to ask. Reports of low taxes and profits heading offshore fuelled a cynicism that has continued to today.

In response, the American mining company rolled out a series of commercials, all ending with the slogan "Utah – We're Backing Australia." There was a lot of finger-wagging from Taylor, an Aussie who'd made it in Hollywood. Back on the coal conveyor:

> You know, they tell me over the last decade hundreds of millions of
> dollars have been spent up here. The important thing for Australia
> is that every time Utah spends a dollar, it inspires other industries

to spend four dollars. I learnt there's enough coal up here to last for hundreds of years [*Taylor is walking through some tall grass, casually picking off some seeds*] so development can continue to grow, the towns are going to grow larger, and the jobs are going to be increasing.

[*Finger wags*]

Utah is going to spend a lot more money in Australia, as will other industries.

[*Shot of sunset*]

Utah believes in backing Australia.

Déjà vu, anyone? Today those wagging their finger may have changed, but the message is the same. Will Australians ever be able to erase the memory of the billionaires' protest in 2010 that saw Gina Rinehart, Australia's richest person, stand on the back of a flatbed truck in Perth and yell until she was hoarse into a megaphone, "Axe the tax!"

Or Palmer's insight to the ABC's Sarah Ferguson as he flew his jet to Canberra to prosecute his campaign against Prime Minister Kevin Rudd's "super" mining tax. "If you have a private plane," he said, "it means you can take your workers with you wherever you go."

Xstrata did the old "suspend two projects" trick, saying the tax would threaten 3000 jobs, and joined Rio Tinto and BHP behind the scenes to launch a relentless lobbying operation, employing Tony Mitchelmore, who ran the Kevin07 campaign, and Geoff Walsh, a former national secretary of the ALP and ex-staffer to prime ministers Bob Hawke and Paul Keating. Through the Minerals Council of Australia, the big three poured $22 million into a television and print campaign.

We all know the end of that story; it is etched into a generation's political memory. Overnight, Rudd was felled by Julia Gillard, who negotiated an "agreement" with mining companies that saw billions shaved off the no-longer-super tax. The boom kept booming and Rinehart became the richest woman in the world. The finger-wagging went global after that. Paul Cleary described Rio Tinto's Tom Albanese saying to a group of mining executives in London that the Australian experience "should send a salutary message to governments around the world. Governments should 'learn a lesson' from the episode, he declared."

Australian politicians certainly learnt their lesson, as it became clear who wielded the stick. If mining companies could help topple a prime minister with a 4 per cent lead in the polls, then surely they could help make a prime minister out of a politician most Australians didn't like? The Opposition leader, Tony Abbott, put his nose to the ground. He declared he would "axe the tax," miserly though it now was, and set about wooing the mining sector. He got on especially well with Paul Marks, the executive chairman of mining firm Nimrod Resources. At a late-2012 dinner in a parliamentary suite, Abbott and Liberal Party colleagues, such as Ian Macfarlane (who went on to become resources minister in Abbott's cabinet and is now head of the Queensland Resources Council), are reported to have assured Paul Marks of their policy to kill off the mining tax. In the *Wall Street Journal*, Daniel Stacey described the "intimate dinner" revealed to him by an attendee. A project of Marks, wrote Stacey, needed "funding and government permits."

> Tony Abbott and colleagues urged a representative of South Africa's Investec Bank to help raise up to A$10 million (US$7.8 million) to advance the project in eastern Australia, according to an attendee. As glasses were replenished with white wine, party members also discussed ways to speed up project approvals ...

Within a year, Marks became the Liberal Party's biggest donor. In 2015, now as prime minister, Tony Abbott chartered an RAAF jet to attend Marks's birthday party in Melbourne, travelling from Brisbane to Sydney for Mike Baird's election launch and then on to Melbourne. He claimed a travel allowance for the trip. It was a busy day. The Coalition had its priorities in place. "Coal is good for humanity," Abbott said on cue as America's Peabody Energy, the world's biggest coal company, began to roll out its energy poverty campaign.

On Peabody's website you can watch an interview with Brendan Pearson of the Minerals Council of Australia discussing Australia's devotion to the cause of reducing energy poverty in the poorest parts of the world. Peabody's vice-president at the time, Fred Palmer, is good at campaigns. He was one of the brains in 1990 behind a decade-long push by a coalition of coal companies to "Reposition global warming as theory (not fact)."

In February 2015 American coal lobbyist Bernie Delaney emailed Australian ambassador Sam Gerovich at the Department of Foreign Affairs and Trade in Canberra on behalf of two Peabody Energy executives, who were planning an Australian visit. Delaney wanted a meeting for them with the ambassador and "relevant colleagues" to "discuss US moves to have the OECD enact a policy guidance document which restricts funding for coal-fired power generation projects." Ten minutes later, Gerovich emailed back with a slot on Peabody's requested date.

The next day, Delaney made contact with another DFAT official, Brendon Hammer, asking for a meeting. Hammer replied, "Always happy to see you and Peabody. Hope the new year has begun well for you." The emails are a glimpse into a concerted campaign run by Peabody Energy to undermine global talks – and the Coalition responded with gusto. (These emails were released after a canny Freedom of Information request and reported by Graham Readfearn in the *Guardian*.)

In November the same year, two Coalition ministers, Josh Frydenberg and Andrew Robb, became key opponents of efforts by the United States, Japan, France, Germany and others to stop the financing of all but the cleanest, most efficient coal-fired power plants, that is the "ultra-super-critical" ones. Australia pushed that OECD countries should be allowed to fund lower-standard power stations and rejected a clause that would require project developers to consider cleaner, renewable alternatives. After almost two weeks of negotiations, Australia, joined by South Korea, finally compromised, signing only if a clause was added to allow funding for less efficient coal-fired power stations in the poorest countries. Reporting the international agreement, the UK's *Financial Times* wrote that sources close to the talks said that while it was monumental, it "would have been stronger if it were not for resistance from Australia and South Korea."

Also in 2015, under Coalition guidance, Australia pledged $930 million and joined the new Asian Infrastructure Investment Bank. It had been agreed by its founding members that it should be a "green" bank. Yet as the bank developed its energy strategy, Australia lobbied for greater investment in coal plants and raised concerns too much emphasis was being put on renewable energy projects. There was a certain irony that Australian lobbyists were so concerned about energy poverty while the Coalition had cut $1.1 billion from foreign aid.

*

Can coal exports be benevolent? Australia started exporting thermal coal for electricity in the 1980s and for about seven years it was an uncrowded market. But when other countries started to get in the game, a plan was devised with a moral twist: to fund coal-fired power stations in developing countries. This may be the modern origin story of coal evangelism. It was missionary work – with cash. At times it truly was a form of aid, as humanitarian agencies such as AusAid funded foreign coalmines. But while the zeal has remained – think of the hokey Carnival of Coal in the NSW parliament, where politicians and coal executives stood around a lump of coal, shouting "Hip-hip-hooray" – the mission has started to rot.

"Gautam Adani's dream to light India's darkened nights" is the headline of Michael McKenna's investigation in the *Australian*. Vividly written, it tells the story of fifty-five villagers without power, people who fear being left behind as India emerges to become an economic titan. One villager told McKenna, "Every time there is an election the politicians tell us they will bring electricity, a small solar plant, but then they forget."

"It was the same sort of promise to modernise India that gave the spark to the 2014 campaign of Narendra Modi ahead of his election as Prime Minister," writes McKenna. "Early last year, Modi promised in his

Independence Day speech to provide electricity to the estimated 18,000 villages still in the dark – with a combined population of 300 million people – by 2019."

This is the challenge India faces. It is very real. In 2014 the World Bank cited India as the world's largest unelectrified population. In some remote areas, the horizon is lined with bare utility poles, which have never been threaded with wires. In mud-brick homes, cooking with wood, kerosene, dung, whatever can be found, thousands of Indians, mostly women and children, are poisoned and die prematurely each year. Children trying to do homework beside kerosene lamps have tears running down their cheeks, their eyes sore. These same lamps are often knocked over and homes – and sometimes families – are set alight in seconds. In the cities, intricate slums house millions of people, mostly with zero electricity but for a tangled knot of black cable tricked up from a main powerline and eked out of the grid.

Coal-fired electricity is expected to bring millions out of poverty, but it is a double-edged sword. In India, rapid industrialisation has brought with it thick pollution. In 2013, a study by two scientists, Sarath Guttikunda and Puja Jawahar of Urban Emissions, an air pollution research firm, found that between 80,000 and 120,000 deaths are caused each year in India by coal-related emissions. The detailed study, funded by the Conservation Action Trust and Greenpeace, also found 20 million new diagnoses of asthma and an estimated cost of $3.3–$4.6 billion each year in hospitals and health spending. The complexity of the problem India is facing is enormous – and it is a moral one for the Western world. In Gina Rinehart's speech in New Delhi this year, the mining mogul sympathised with the Indian people:

> Growing up in the outback in the north of West Australia in remote and hot areas, without electricity, but supplied by generators operating approximately two hours or less a day, I have some understanding of life without electricity.

She also has some sympathy for her yet to be unlocked coal deposits in the Galilee Basin. India was meant to be the next boom. In 2010, when Gautam Adani bought his Carmichael coal deposits, India had a supply deficit of almost 15 per cent, and that was just for those hooked up to the grid. At the time, his plans aligned with government policy: coal was India's future. India's outlook has since shifted. Last year, India's energy minister, Piyush Goyal, announced plans to see almost 60 per cent of the country's electricity derived from non-fossil fuels by 2027. He also announced that apart from those already being built, India would not be approving new coal-fired power stations. Adani has shifted accordingly. McKenna compares the lengthy assessment process for Adani's proposed mine in Australia to the company's swift construction of one of the world's largest solar plants in the south of India:

> The world's largest single location solar plant, it took just eight months to build: 8000 labourers worked day and night to clear the fields, erect the panels and converters and put lines out to the grid. "We would bring in the components at night and lay them out and then put it together during the day," project manager KS Nagendra says. "It was a big job and no one got hurt. Some got bitten by snakes and scorpions, but they are everywhere here. We would have done it quicker but we were slowed by the monsoon."

Solar power is a big part of India's energy plans. But, McKenna adds, "expansion of electricity will also remain hugely dependent on coal. The International Energy Agency estimates the fossil fuel will make up 60 per cent of India's supply until 2020." Others disagree – and it is true the International Energy Agency has consistently underrated the renewables sector, repeatedly having to adjust forecasts on solar costs and installation as prices tumble due to China's mammoth shift to renewables.

In 2008 New Delhi's Energy and Resources Institute launched the "Lighting a Billion Lives" program, rolling out decentralised grids using photovoltaic solar and LED lights to more than 3000 villages in India and

Africa. It is one of many such initiatives throughout the world, some of which train locals to maintain solar microgrids. In 2014 the Nobel Prize in Physics was awarded to three Japanese scientists whose discoveries in LED lighting are transforming the developing world. In tandem with solar panels, they have provided millions of people in Africa and Asia with electricity for the first time in their lives. The process is called technological "leapfrogging." Advocates point to the rapid transformation mobile phones have brought for the millions who were unable to access traditional phone lines. Now almost everyone has a mobile phone. E.A.S. Sarma, a former secretary of India's Ministry of Power, has said that India's energy poverty will not be alleviated by more coal and that the "huge social costs outweigh the perceived benefits." In the *Guardian* two years ago, Sarma wrote that the claim by Adani and Australian politicians that "100 million will be lifted out of energy poverty by a new wave of coal exports … reveals a deep lack of understanding of the real situation in India."

The poverty of millions without access to electricity is a complex and tragic problem. Such problems have intricate histories, and for their victims they may also mean an uncertain future in the face of rising temperatures.

Perhaps we can judge by their actions the sincerity of those in Australian politics gripped by a newfound commitment to reduce global poverty. The ferocity and grit with which the Australian delegation fought a clause in the proposed OECD agreement that would require project developers at least to consider a clean energy alternative sums that up perfectly.

*

I thought a lot about the crown-of-thorns starfish while driving through the Bowen Basin. Great big spiky things, brown and black, sometimes a velvet red, expelling their stomach over the coral, so hungry are they. And yet they belong; they've been a part of the reef's ecosystem for eons. But in the past few decades something glitched – marine scientists still don't know for sure what that glitch was, or is. It could be nitrogen run-off from farms, sediment from land clearing, fisheries destroying natural predators – perhaps all three. At night, divers will tell you, a reef is either the most exquisite and luminous organism there is, unfurling tentacles, sleek grey sharks, manta rays. In their torchlight, divers might see worlds of plankton, swirling specks of varying colours and the clams, corals, anemones reaching out to catch them. Or they'll see hundreds of crown-of-thorns starfish, even stacked one on top of the other, trying to eat the coral underneath, growing in size before the night is done. People, volunteers mostly, can collect a thousand in a single day while the starfish sleep under what remains of the coral. I am not sure what plan B is for the starfish if they eat all the coral. I am not sure starfish think that far ahead.

Coalmines are not foreign either. Well, they're foreign-owned, yes, but they are a part of the Australian story.

At one point as I continued down the highway, I felt the car tremble. I registered it and kept driving. It was only much later, when the dusty green brigalow and tall bottle trees started to appear, that I realised it must have been a blast.

In March this year, the ABC announced a significant restructure that would create eighty new jobs in the regions. It is an overdue, necessary development. What has been happening out on the land, to the land, has been largely unregistered by the nation. In the name of coal and coal-seam gas, towns have been erased, vast tracts of prime farmland destroyed, rivers diverted, more than 30,000 sacred sites wiped off the face of the Earth. Widely respected journalist Paul Cleary has documented the fate of thousands of rural Australians in the shadow of the recent resources boom, tracing a sadness that has seen farmers, graziers and their families – entire regional communities, in a bizarre echo of their Aboriginal predecessors – lament the loss of country.

"So great, so far-reaching and so uncontrolled is the resources boom," wrote Cleary, "that it appears to be driving the biggest forcible transfer of land on this continent since the first wave of white settlement."

In Queensland, much of this is done under the guise of helping to develop the north. Who can forget the fluttery voice of Gina Rinehart warning Australians against complacency, reminding us that "Africans want to work and its workers are willing to work for less than $2 per day"? In the same 2012 video, Rinehart floated an idea close to her heart: to turn northern Australia into a "special economic zone" with lower taxes, fewer regulations and – as revealed later – weakened native-title rights.

Rinehart's dream involved a lot of networking. In 2011 she flew three Coalition ministers, including the now deputy prime minister, Barnaby Joyce, and the now foreign minister, Julie Bishop, in her jet to India to attend the wedding of Indian billionaire G.V. Krishna Reddy's grand-daughter, an affair boasting 10,000 guests. Julie Bishop later claimed a $3445 flight home. Rinehart didn't fly her back to Australia – and the trip, Bishop declared, was a "study tour" in which she had meetings with various Indian companies. Three of her meetings, she reported, had occurred on the day of the wedding, one with the bride's grandfather, Rinehart's business partner, Krishna Reddy. Bishop was later asked to elaborate. "The wedding started at 6 p.m. on the Saturday night," she said. "The meetings were held prior to the wedding."

Barnaby Joyce claimed $3600 for his and his wife's flights to Perth and a further $350 for his day in Perth before the couple boarded the private jet, as he had some meetings in Perth, of course. He also claimed their $5500 flight home from Kuala Lumpur (the private jet dropped them off there after the wedding). His day in Malaysia was also a "study tour." He met with "officials" and in an obligatory report shared this gem: "Malaysia has recently experienced high levels of economic growth, which has created urban cities comparable in wealth to cities in developed countries." He also noted that Kuala Lumpur's "substantial freeways ... would look quite in place in an Australian major capital city." Presumably this insight was gleaned on the way to the airport. Three months after the wedding, Krishna Reddy sealed the deal with Rinehart, his GVK conglomerate buying a major stake in her Galilee Basin mine for $US1.26 billion.

The next step was obtaining state approval. In May 2012 Queensland officials told their counterparts in the federal environment department that the state was set to give conditional approval to the GVK Hancock coalmine, even though, Paul Cleary writes, "work on assessing the project's full impact under federal law was still some way off." An official wrote to a federal counterpart in a leaked email:

> I expect that Hancock will be lobbying heavily to obtain their approval from you once our report is finalized; they have had a direct line to the new government and the Co-ordinator-General here. On Tuesday [22 May] they came in with 22 experts to "discuss" the proposed conditions, 48 hours before the report was supposed to be finished.

In 2015, when the Coalition launched the Northern Australia Infrastructure Facility – from which Adani expects to receive a billion-dollar loan – many noted it was eerily on point with much of Rinehart's vision for northern Australia. Focusing predominantly on mining and irrigating farms, the fund seems to ignore a central reality of northern Australia. By 2040, half of the population will be Indigenous, and yet the fund is – as Yawuru leader Peter Yu has said – "practically silent" on economies that are proving successful in Indigenous communities: land management, carbon sequestration, conservation and ecotourism.

In February this year, Matt Canavan, Minister for Resources and Northern Australia, said Adani was being exposed to "unfair scrutiny." Revelations the conglomerate is being investigated in India for alleged tax fraud and money laundering were followed by the Adani Brief, a compilation of court records put out by the legal group Environmental Justice Australia. Of particular note are Indian tribunal findings:

> In August 2016, the National Green Tribunal [India's specialised environmental court], fined AEL [Adani Enterprises Limited] nearly AU$1 million for its role in chartering an unseaworthy ship to transport coal. The tribunal found that the ship had sunk, spilling oil and over 60,000 tons of coal that destroyed mangroves and polluted beaches. The tribunal criticised AEL's failure to clean up the spill for more than five years.
>
> In January 2016, the National Green Tribunal fined Adani Hazira Port Private … almost AU$5 million for undertaking development works at its port in Hajira, India, without an environmental permit. The tribunal found that these works destroyed mangroves and impeded the fishing activities of local communities by interfering with their access to the river and ocean. The tribunal criticised the company for having an "irresponsible attitude" and for failing to care about any "adverse impact [of its development] on [the] environment."

In 2015 the ABC's Mark Willacy revealed that Adani's Australian CEO, Jeyakumar Janakaraj, had been in charge of a Zambian copper mine when it discharged dangerous contaminants into a major river.

MARK WILLACY: These 2010 Zambian court documents show that KCM [Konkola Copper Mines] pleaded guilty to four charges of contaminating the river with toxic pollutants and of wilfully failing to report it. The company was fined the equivalent of several thousand dollars. A few months later, there were Zambian media reports that the company had again contaminated the river.

In the London High Court, Janakaraj has been drawn into a claim against KCM by 1800 Zambians, who allege the mine poisoned their water and their health. Of Janakaraj's failure to disclose this information (a legal requirement for Australian assessment), the federal environment department told the ABC that it was most likely due to a "mistake" and that "the omission did not result in environmental harm."

Matt Canavan is right. The level of scrutiny now applied to Adani is in harsh contrast to the lack of scrutiny given to almost every other mining and gas company in this country over the past two decades.

Let's consider Linc Energy, the company which sold Adani its Carmichael deposit. Last year Linc went into voluntary administration, then liquidation – a manoeuvre, critics suggest, to avoid facing a criminal trial for five charges of environmental damage after its coal-gas facility in Chinchilla contaminated vast tracts of Darling Downs farmland. An expert report has determined that the damage to at least twenty farms is irreversible. Richard Guilliatt reported in the *Weekend Australian* in June 2016 that eighty-odd farming families within 320 square kilometres of the facility have been put in an "excavation caution zone." He wrote:

In the 16 months since then, they've become a lot more enlightened. They've learnt that Linc Energy stands accused of fracturing the rock beneath their land and releasing toxic chemicals into the soil, air and groundwater over a six-year period. They've read that Linc's workers were told to cover up the contamination and drink milk to protect themselves. They've been told that digging a hole in a paddock might release "potentially explosive and/or toxic and/or asphyxiating mixtures of gases." They've heard the Queensland environment minister, Steven Miles, describe it as "the biggest pollution event probably in Queensland's history."

It all sounds like a rather surprising development, except it wasn't. Guilliatt dissected more than a decade of warnings to the Queensland government from insiders, as well as serious concerns and complaints from the Chinchilla community, all of which were ignored. Hyped as "clean coal," the Linc Energy facility's method was to ignite the underground coal seam to generate gas. In the early 2000s Queensland's Labor government under Peter Beattie approved the facility and categorised it low-risk, so that it did not require an Environment Impact Assessment – in spite of CS Energy, a state-owned energy provider, pulling out of the joint venture with Linc due to significant concerns about the technology. Richard Cottee, chief executive of CS Energy, told Guilliatt he had pulled the plug because Linc couldn't explain what had happened to the carcinogenic benzene and toluene contaminants it produced. "We knew they had to be produced, but where were they?"

No baseline studies were done on groundwater and soil quality before the facility was launched – a common feature of the state's rapid rollout of mines, mine expansions and coal-seam gas drilling, the latter of which has seen over 5000 of an approved 40,000 wells drilled across what locals now call the "pincushion" state. Wells have been sunk on land owned by traditional owners, graziers and farmers, who for the most part have had little choice but to open their gates to the gas companies. State law, writes Cleary, allows gas companies access to land after a minimum negotiation period "of just twenty business days."

In February this year, the Darling Downs contamination zone was widened to the south for another twenty-five kilometres. As for Linc Energy's Peter Bond, in response to the Queensland government's pursuit of him for a $5 million guarantee to the estimated $28 million rehabilitation of the Linc Energy facility, he told gathered reporters in disgust, "Queensland is closed for business."

*

I was driving to the Galilee Basin. The dirt turned red, locusts smacked against the windscreen, and I started to feel good. Sure, the land was patchy, the trees too thin, grass scoured, but at least *it was there*. I stopped in Alpha, a little town of 500 people. I had driven nearly 700 kilometres from the coast. In the public toilet block, someone had woven fake flowers around the mirror above the sink. There was a plastic bottle of hand soap sitting on a cream doily, and a clean folded napkin. In ballpoint pen, someone had written on the napkin, "Thank you! What a lovely little town." I kept driving, up to Jericho, and then onto a dirt road. By now my tiny white jellybean was caked in red dirt and locusts. It flew over the corrugations like a package about to fall apart. Long-legged birds with red heads carefully picked their way over a small swamp. A couple of skinny kangaroos bolted. After an hour, I got to the turn-off to Speculation, a cattle station run by Bruce and Annette Currie. Their neighbour is GVK Hancock. Of the nine mega-mines proposed in the Galilee Basin, two of them, GVK Hancock's Alpha and Kevin's Corner, a colossal sprawl of open-cut and underground mines, will be nudged up next to the Curries' property.

I turned and drove the jellybean over ten or so kilometres of track, steering carefully around chunks of rock, crickets chirruping and flying into my hair. Finally there were a couple of buildings, dusty yellow Maremma dogs in the shade of a truck, a horse, a row of cattle dogs in *Footrot Flats*–style kennels. The gate was tricky, a tangle of wire connected by a star picket. I couldn't open it; sweat was pouring off me. I learnt that standing still out here, feet in sandals and legs bare, was a bad idea. Out of red cracked dirt hundreds of ants got busy and I started to shake my legs as they sprawled like capillaries over my skin. It took three trips of running back to the safety of the bonnet and returning to the gate to get it open, another five on the other side to close it.

Past the gate, four goats jostled, oblong eyes sizing me up. They surrounded the jellybean, sniffing it. One climbed up on the bonnet and I groaned. Ahead was a ramshackle building, some of the doors mismatched. Next to the front door was a two-way radio. Two voices crackled out of it, talking to each other. Bruce Currie was mustering his Brahman cattle. He'd pegged a note to the door in cursive hand. Inside was a kitchen, a table with mismatched chairs and loaded up with books and stacks of paper. On the wall over the desk was a farmer's filing system – a row of bulldog clips under headings such as "Broken." I made a cup of tea and headed back outside. A tiny oasis, the place was croaking with frogs. They splatted past me, one doing a magnificent backflip to catch a fly. The toilet was outside. When I spun the roll of paper, an emerald frog snug inside jumped out. I squealed. On a line strung across the veranda were six pairs of faded blue jeans. I blew on the tea, legs folded under my bum so the ants couldn't get me.

In 2013 and again in 2015, Bruce Currie took GVK Hancock to the Land Court over the two mines' impact on groundwater. The Galilee Basin is connected to the Desert Uplands region and the Great Dividing Range. Seven rivers make their way across the basin. Some then flow northeast to the Great Barrier Reef, the others inland to Lake Eyre. Under the earth here, as in most places, is a strata of local aquifers – which the individual mega-mines will, in mining-speak, "de-water." These aquifers are connected to the Great Artesian Basin, the only source of fresh water in inland Australia. The vast basin lies beneath 22 per cent of Australia, from the northern tip of Queensland stretching across into the Northern Territory, down to Dubbo in New South Wales and past Coober Pedy in South Australia. Farmers, industry, landowners and about 200 towns tap into this intricate water source, which, as the Great Artesian Basin Coordinating Committee explains:

> fell as rain 1 to 2 million years ago and percolated into a giant underground basin … It is not a lake at all; it is solid rock, with water

stored in the pores between the coarse grains of sand in vast sand-stone sheets. And the Basin's water is connected to aboveground life-cycles, received as rain, and emerging in springs, streams and bores.

Mining companies are required to obtain licences to take water as part of their project assessments, but it is a fundamentally flawed process. State governments have a clear conflict of interest, with royalties and revenue at stake. It would not be a stretch to suggest government departments overseeing applications are pressured to "work backwards" – that is, to begin with an approval and ensure the paperwork aligns accordingly, per-haps with the help of a company's own consultants. In the Bowen Basin and now here in the Galilee, the mining companies have negotiated make good agreements with landowners to compensate for when the mines pump aquifers dry, as has happened in the past – and which may be irre-versible. In any case, if ever put into effect, a make good agreement means that a cattle station out here is as good as dead. Out here, the sky is not something to rely on.

As the sky began to turn a deeper shade of blue, streaked with pink, I heard the bellowing of cows. I stood up to see the Brahmans coming in and two people on horses, calling out to each other as three dogs circled and yipped, nudging the herd over the rise and into a holding yard.

It is a complex thing, this meeting of worlds – climate change and cows, the reef and graziers. Bruce and Annette's Brahmans are at the beginning of a chain of events that will ultimately see them be loaded onto a ship and killed overseas. A conclusion I have seen with my own eyes, and which I cannot abide. But Bruce, tall, wiry, with sunned skin, bristles of hair and thoughtful green eyes, is a man open to ideas; there was no aggression or salesmanship as we talked. He and his daughter, Renee, hosed down the horses and dogs as the Brahmans quietened.

Bruce, now fifty-six, grew up on the land, a cropping farm in the Bowen Basin. His father died when he was in high school and he shared the load with his mum. But when the huge open-cut coalmines started to appear a decade ago, he and his wife sold up and they came out here. "Fat to the fire," he chuckled. After going back and forth with GVK Hancock over the wording of its make good agreement, he and Annette found they were racking up thousands of dollars in legal advice just to understand the documents. The wording was "opaque," Bruce said, and he couldn't get it any clearer. "They kept saying others had signed it."

Why? I asked. Bruce shrugged. About to retire anyway, kids aren't going to take over the land. It's too hard, too expensive to fight, and you could end up losing everything. They also signed the confidentiality clause attached to the agreement. "It divides you from your neighbours." But still, he and Annette wouldn't sign. For Bruce, the sticking point was that the onus of proof was on them if their water ran dry. They decided to fight. "We could ignore it, let it devour us," he said. He waved his hand out at the land. "In the papers, they called us greenies, just because we wanted to know what would happen to the water on our land. We're farmers. Every farmer knows without water, they're gone." Bruce represented himself, working through the night, studying the law, a highlighter in hand. In the first court hearing, Bruce was joined by two other landowners, including Paola Cassoni, who had, in 2000, bought 8000 hectares of significant bushland to protect it from land clearing. The federal government kicked in $300,000 to support the purchase and it was approved as the Bimblebox Nature Refuge three years later. It is now gazetted to be mined as part of Clive Palmer's China First coalmine. In response to petitioning to protect it, Palmer said contemptuously, "There's nothing there of any environmental significance. If people want to dress up as kangaroos and koalas, they can do it." There were 9000 jobs, he went on, and children who needed Christmas presents. "Your children are more important than any wombat, I can tell you that."

But the Land Court sided with Bruce Currie, who lugged a plastic container of files into the hearing every day, and the two other landowners. Of the proposed Alpha mine, the presiding judge expressed a "lack of confidence" in GVK Hancock's groundwater evidence. A happy Bruce told reporters, "Well, if self-represented non-lawyers can poke holes in a mining company's environmental assessment, just imagine the quality of the work that mining companies want us to take as gospel!" The court recommended the mine should not go ahead, but if it does, there should be improved monitoring of bores, and make good agreements. However, those recommendations aren't binding on the state government.

For Bruce, it was an exhausting process of taking one mega-mine at a time to court. In 2015 he took on Kevin's Corner, GVK Hancock's other mine. The presiding judge has still not made a decision. In turn, Bruce has discovered a world of law, governance, hydrology and climate change. He has drilled down into everything he can find, reading voraciously. He is a man who doesn't like not knowing things. In court, he told me, there were times when he could not understand what GVK Hancock's experts were saying. "I had to just let it pass. It was embarrassing." In turn, the Coalition raged against him and other court actions, with Prime Minister Tony Abbott labelling these "sabotage" and instructing the attorney-general, George Brandis, to rewrite the legislation to gut sections of the Environment Protection Biodiversity Conservation Act that allow for a variety of appeals. To date the repeals have not come into effect, but Malcolm Turnbull has pledged support for the changes. The Queensland government has made an amendment to exclude the Adani mine from public appeals such as Currie's and other landowners' with regard to its water licences, while less than two months ago, in a midnight session – while Cyclone Debbie raged across the state – the parliament granted Adani an unlimited water licence that will see the mega-mine through to 2077.

Farmers and graziers are well aware of the bitter irony. In the past twenty years, use of the Great Artesian Basin has been extensively reformed, with over 650 bores capped, saving around 200,000 megalitres each year, and landowners made subject to stricter regulation. This reform was in response to a significant decline in pressure that has seen springs in Queensland and South Australia dry up. Ellen Moon, a geochemist, wrote in the *Conversation* that, "Flows from artesian bores are now roughly half what they were in 1915. Since then, the water level in some bores has fallen by as much as 80 metres, and a third of bores have stopped flowing altogether." Scientists estimate that the basin is refilled slowly over millions of years – between 0.5 and 10 millimetres annually – but the fear is not that it will run dry, but rather that the pressure that surges the water upwards in a bore will continue to diminish.

Yet, despite awareness of this, approvals for mines and coal-seam gas wells – which in Queensland account for 22 per cent of the total water extracted but are exempt from licences altogether – are being waved through. Coalmines use a lot of water. In its Environment Impact Statement, Adani forecasts it will need 12,000 megalitres each year. This is about thirteen Olympic swimming pools per day and will be used to manage spot fires, dampen coal dust and for use in processing. After operations have ceased, these open voids drain the watertable for generations to come. The concern is not only that the mines will "de-water" nearby bores – a possibility acknowledged by mining proponents – but also that they will increase the risk of salinity and damage important recharge springs. Attached to Adani's "unlimited water licence" – an oxymoronic term – is the requirement that Adani will regularly monitor water levels, which will, in turn, be monitored by the state. This has instilled little trust. The state has largely failed to monitor the rapid rollout of coal-seam gas wells – which have seen many landowners claim their bores have not only run dry, but are bubbling methane instead. A fracking incident in 2009 that saw the company QGC, in its words, "unintentionally provide a route for water into the aquifer," giving rise to fears that chemicals had entered the basin, was revealed not by the state, but by farmers.

For rural Queenslanders the Land Court is the only remaining place where they can exercise their right to water – albeit through a lengthy, exhaustive process that few are able to undertake. But with Adani's water approvals now exempt from the judicial system, one wonders how much weight the court's rulings carry, both now and in the future.

Tony Abbott's claim of "sabotage" had an element of truth. In that this is the way such battles are fought – in the courts, at company AGMs, with divestments and boycotts, and, finally, in the streets. There are the usual, sometimes uneasy alliances of farmers, conservationists, fishing and tourism workers, and traditional owners. When the judicial process has not gone its way, the Coalition's response has been to attempt to change legislation rather than consider how to give effect to rulings. In the name of the Adani mine, attempts have been made to amend and gut native title and the environment act, while the state government has cut off avenues of appeal. In the name of digging up the Galilee Basin, the Coalition has put forward measures to "protect democracy" from foreign donations, such as the funding offered to the Wangan and Jagalingou people traced back to the United States.

There is so much to learn, to discover, not just in the books stacked on Bruce's kitchen table, but outside, in the night, a sky full of pinpricked stars. There are many mysteries remaining about the ancient water source underneath the land; so too about the coral atolls that form a kind of labyrinth off the coast. But on all that we do know, the Queensland and Coalition governments have thrown all of us – not just a bunch of cattle stations, cropping farms and native-title owners – under a juggernaut of enormous driverless trucks.

*

In Emerald, the man at the reception hovers, waits for his wife to book me in. I've spent an hour at the car wash, cleaning up the signs that I took the hire car off-road. He's in his seventies, with watery blue eyes. He tells me he ran a farm twenty-seven years ago – grain.

"Hard work," I say, and he smiles.

"Good hard work. I liked it."

"You miss it?"

He nods. His wife is now efficiently typing in my details on a computer. "Was my wife's idea to run a motel," he adds. A barely perceptible, chilly ripple goes through her. She looks up and smiles at me. Asks where I've driven from and I tell her from out Jericho way. It's a long trip. She's confused. "You're a nurse?"

I tell them I'm a writer. The wife's eyes light up. She wants to know if I write longhand and then type it up. "Is it a book?" she asks. I'm stuck. I don't really want to say because it feels like I'm setting them up. "It's just journalism," I say. "About the mining sector." She tilts her head at me, then looks down and finishes typing in my details. It's been tough, her husband says, the downturn. "I suppose too many people opened up hotels and bought properties." His wife stiffens; I can read her body language even if he can't – or won't. *Darling, shut up.* He straightens, thrusting his neck forward. "My personal opinion is they need to get this Adani mine started and all the economy going again."

She puts my key on the counter and I stare at it. I struggle to compose the thing inside me that wants to leap over the desk and throttle him.

He meets me at my room, carrying a basket with a towel in it and a small jug of milk. I open the door to my room and sigh. There are no fly-screens. No fan. He offers to turn the air-conditioning on for me. I smile. No thanks. He wants to talk. But I'm tired. I miss the frogs. Poor bastard, he misses his farm. I miss the old dogs pushing their heads into my hands. The goats jostling at the gate. Bruce with his thoughtful manner and his "Hurry up, Dad, the cows are waiting" daughter. The red earth and dusty brigalow. The bottle trees, some of their heads lopped off by cyclones.

Then, in South Australia, the lights went out.

The people of Port Augusta came out to watch the crop duster. From the small plane's belly, pastel green clouds belched over the ash dam of the Northern Power Station, a hulk of cement and rust on the saltpan landscape. It had chugged along for over thirty years, while behind it the Playford B station had been in the game for over fifty, powered by brown coal from the Leigh Creek coalmine. That November day, the clouds descended to form a sticky lurid resin atop a flat dam about the size of Adelaide's CBD. The coating stopped the ash from blowing over Port Augusta. Back when the power stations were running, they had pumped water in from the Spencer Gulf to cool their boilers and then poured the slurry of seawater over the ash to form a salty seal.

There is now a distinct fishy smell and a flush of midges as you drive over the saltpan into town, another result of the closure. When complaints over the stink made the news, the owner of the stations, Alinta Energy, told reporters it had offered the council the option of keeping the pumps running, but the proposal was turned down. A local councillor, Lisa Ludsman, told me wryly, "They neglected to mention it would be in exchange for a third of the ash dam's liability."

It was last year, in May, when a dwindling workforce at the Northern twisted down the dials in the control room and watched the screens flatline. Playford B had been mothballed four years earlier, squatting in the background, ready to creak into action if needed. It was eerie shutting down, one worker recalled to me, for a coal-powered station is like a heart. They are rarely, if ever, turned off in their lifetime – always humming, boiling, burning, cooling, fed from a stockpile endlessly replenished by the back and forth of coal trains. Thus the term "baseload," a recent addition to the political lexicon intended to paint these ceaselessly functioning old power stations as a reliable father figure, always there in times of need.

Or else as an inefficient and outdated technicality. In South Australia, wind turbines and photovoltaic solar panels now provide 40 per cent of the state's electricity. It was this and the nearby Leigh Creek mine being scoured completely of coal that forced Alinta to close the Northern and Playford stations. In a significantly altered electric landscape, their mandate had shrunk. Required only in times of peak demand, these hulking creatures were too costly to run constantly. Coal was no longer cheap.

"It is a noble art, making electricity," Gary Rowbottom told me. Rowbottom lives with his wife in a blonde-brick home in a neat street overlooking Port Augusta. For seventeen years he worked at the Northern and when the station's 80-metre-tall smokestack was to be detonated, he set himself up on the hill and filmed it with subdued sadness. Sadder still, he said, was when one of the demolition team sent him a video of the turbine being cut into sections with blowtorches. As part of his job, Rowbottom had spent a lot of time and mental acrobatics trying to nut out how to upgrade the turbine blades. "Everyone had been privatised by then, so no one shared information anymore, so I was on my own with that problem." It was a complex challenge he enjoyed solving with the help of two engineers from Sydney and Hobart. "I had to go up to Kalgoorlie to meet them between their twelve-hour shifts." Rowbottom's father had worked at the Playford power station and his son, now a mechanical engineer, joined him briefly at the Northern. You could surmise that power stations are in his blood, but it would be more accurate to say electricity is his passion.

A decade ago Rowbottom looked at the horizon and knew things were going to change. "I have been reading *National Geographic* for twenty years," he explained. "I had a fair understanding of climate change and thought, shit, we're lucky. We have time to fix this." So when, in 2012, an alliance of locals, power-station workers, councillors, environmental groups and unions formed Repower Port Augusta, it was natural for Rowbottom to be at the centre of it. The idea was to develop a vision for the town post-coal. At work, he became "that guy" obsessing over the future. But the future kept coming a little closer. The town swung behind the group, largely due to the ferocity of the mayor, the late Nancy Joy Baluch, who led Port Augusta for the best part of thirty years, never hiding her loathing of the two power stations at the edge of town. Baluch blamed them for her son's chronic asthma and then for the death of her husband, who had worked at one of them, a non-smoker felled by lung cancer. She was instrumental in advocating for upgrades to lessen the ash and coal dust that blew over the town. In the 1970s, when the local newspaper wrote an editorial raving about the jobs and economic opportunities the state had created by adding an extra boiler to Playford B, Baluch unleashed her fury. In a letter, she challenged the "city fathers, the Government and Health department" to come to the children's hospital and sit with her and her husband beside their seven-year-old son. The ward, she said, was full of children from Port Augusta with chronic asthma.

In 2013, just before Baluch died of breast cancer, the National Library of Australia interviewed her. June Edwards asked why she had decided to run for mayor in 1980, and in her acerbic way Baluch replied, "Because the others were not suitable candidates."

And they weren't.

Baluch was far from perfect. Political rivals, Michael Dulaney wrote in *Griffith Review*, referred to her as that "skinny bitch," while her admirers called her "a multi-dimensional woman." Dulaney summed up the sheer difficulty of pigeonholing her:

> She was a hater of many things: whingers, whiners, bankers, the "bloody do-gooders," political correctness, the media and sports of any kind, which she considered to be a waste of time and a national distraction that would facilitate a peaceful takeover of Australia on Grand Final day by "the Asians." Joy was intolerant of the "soft generation" and an advocate for setting police dogs on Aboriginal people, but also for changing outdated attitudes towards homeless shelters and domestic violence refuges.

But Baluch valued health and this was what set her apart from her political rivals. Health came first, before jobs, before the economy, before GDP. In this she was unlike any other mayor of a coal-dependent town in Australia. When the state government blamed residents' respiratory and bronchial complaints and lung cancer on Port Augusta's higher rate of smoking, Baluch was furious. After a six-month campaign by the council and Doctors for the Environment Australia for the release of SA Health figures, it was revealed that despite smoking rates in the region being only 7 per cent higher, lung cancer rates were double the national average.

By chance, when I drove into Port Augusta, I picked a hotel with a huge portrait of Baluch hanging on the wall. Her daughter, Michelle, wrote down my details. "I remember watching her up there," Michelle later told me, about a speech her eighty-year-old mother gave, still serving as mayor, on the steps of the state's parliament. "She was so sick and frail, and she was shaking her fist and the crowd was roaring." Afterwards, Michelle and councillor Lisa Ludsman helped her into the car, moving her around in the backseat so the sun wasn't beating down on her. It was 2012 and Repower Port Augusta was gathering momentum. Supporters walked from Port Augusta along the 300 kilometres of highway to Adelaide and were met by over 2000 people at Parliament House, where they rallied for a new vision. A solar thermal storage plant – 10,000 mirrors, each the size of half a tennis court, fanned around a 100-metre-plus tower containing molten salt. Reflecting the sun, the mirrors would focus on the top of the tower to heat up the salt to generate electricity.

Port Augusta has, on average, 300 sunny days each year, but even so, molten salt has an added benefit. It can *store* the energy, which is new in a traditional grid where electricity is mostly generated and supplied in tandem. The solar storage means the project can generate power at night, in the rain, when the wind is not blowing and, most importantly, during peak demand. It sounds crazy. But then so does digging up ancient swamps of peat and bog which dinosaurs waded through and huge dragonflies flitted over that have hardened into combustible rock over millions of years.

Gary Rowbottom has dozens of spreadsheets running on his computer. Every few days he checks the monitoring station for temperature and condensation levels at the proposed site of the solar thermal storage plant. He has a spreadsheet dedicated to Australia's carbon budget, another of all the power generators attached to the grid. "My children call me Captain Obvious," he told me, after running through the reams of modelling and data he has collected that show Australia's grid is well overdue for an overhaul and significant investment. For five years, Repower Port Augusta lobbied state and federal governments to take an interest in what it was proposing. It had two companies ready to tender, had the build costed at $650 million – a lot of money but, after all, Tony Abbott had thrown hundreds of millions at his so-called "Green Army," a proportion of which went to projects such as pulling out weeds in urban areas to protect "habitat to the ... threatened Southern Cassowary." But most politicians listened politely with glazed eyes, startling only when Joy Baluch whacked her walking stick on the table.

Still, Repower Port Augusta kept the faith. Workers at the Northern were under the impression they had jobs until 2030, and Rowbottom envisaged having the storage plant up and running by the time the station powered down. He'd be ready to retire and looked forward to watching the town transform, maybe even seeing his son with his young family return to work. But in mid-2015, power-station workers woke to front-page news in the Adelaide *Advertiser* that they would be losing their jobs much sooner than they'd thought. Just the week before, Alinta had assured them the station would not be closing anytime soon.

Alinta's announcement was another domino falling in South Australia as it pushed towards its 50 per cent renewable energy target. In 2013 Engie's Pelican Point plant had reduced its local gas capacity. A year later, AGL's Torrens Island gas plant announced a plan to withdraw half of its capacity. But the Northern closure was even more significant than these, and representatives from both the coal and the renewables sectors noted potential problems. Wind turbines and solar photovoltaics had inched towards taking over half of South Australia's electricity supply, and both sectors knew the state was going out on a limb without a back-up plan.

Coal players pushed for subsidies and lifelines. Prime Minister Turnbull was urged to take up his predecessor's war on renewables, which had already seen more than 3000 jobs lost in three years and large-scale investment grind to a halt.

Renewable energy advocates urged state and federal governments to take the next step and invest in energy storage. There were calls to keep the power station's steam turbines running in order to maintain flow in the electricity grid to smooth out any voltage disruptions.

Repower Port Augusta waved frantically, but federal and state governments just stared straight ahead.

The Australian Energy Market Operator (AEMO) said the power station's exit would not affect the grid's reliability as long as there was sufficient "synchronous" generation in the grid, a stable frequency of power that can ride out and absorb any shocks. It also noted South Australia would be more likely to need to rely on the interconnector that plugged into Victoria's brown coal-fired power stations.

Lisa Ludsman recalls shaking Malcolm Turnbull's hand during the federal election and seeing a flicker of interest as she told him about the tower of molten salt. Independent Nick Xenophon and his team were gaining in the polls and the Coalition indicated it would give priority to funding solar thermal in Port Augusta. After the election, Port Augusta heard not a peep.

Then, in September, the entire state blacked out.

*

It was called the most ferocious storm in half a century, and it approached from the Southern Ocean. On the Bureau of Meteorology radar you could watch it unfurl. There had been preparations: deliveries of sand for people to start sandbagging. Port Pirie, perched on the Spencer Gulf, was warned of huge waves. Winds, it was forecast, could reach speeds of up to 120 kilometres per hour. AEMO, monitoring the electricity grid, decided there would be no need for a "credible contingency" plan. It was a decision that meant AEMO would not be able to intervene and instruct generators to start load-shedding (such as turning off a smelter). Many in the electricity sector later expressed bewilderment at this inaction.

In the city of Adelaide as the storm approached, people were told to go home immediately or risk getting stuck. Traffic quickly gridlocked and the state went dark as the intense low-pressure system rolled over, covering the sky in thick bulbous storm clouds. Winds started up, in some places getting to 100 kilometres, in others 120. The trees were bending, roots like fingers trying to hold on. Lightning flickered and forked, some 80,000 lightning strikes dappled the ground, coastal roads were covered in sea foam and hailstones the size of a fist. Trampolines got tangled in power-lines. The wind farms were providing half of the state's electricity, roaring in their element. Gas plants in the state, such as Torrens Island, put in a little under a quarter, and the rest was met by coal-fired power coming in on the interconnector from Victoria.

At around 4 p.m. a cluster of tornadoes struck the transmission towers in the middle of the state and jerked twenty-two out of the ground. If they were the backbone of the state's grid, this was yanking the spine out of the body. Others were twisted and dangled like broken chicken necks. It was revealed that ElectraNet, the company responsible for the towers, was due to begin maintenance on the infrastructure and had informed the regulator of this in 2013. However, the company said none of the transmission towers damaged in the storms was due for maintenance.

From here a cascade of six-voltage "disturbances" went down the lines. Within seconds, two-thirds of the wind farms disconnected. The interconnector surged and tripped. The state was "islanded" from the national grid, the gas plants went offline and the state blacked out. Traffic lights cut out.

A generator failed at a hospital and patients were transferred. Embryos thawed. A mother and her two children were trapped in a lift after a dental appointment. Like a great big animal, the state closed its millions of eyes, buffeted in the dark.

Tea-light candles were lit.

In what is known as a "black start," the state had to get back onto the grid from scratch. AEMO instructed AGL's Torrens Island gas plant to turn back on. The head of South Australia's Australian Services Union, Joseph Scales, told me the workers were already on it. "Workers at home rang their colleagues at Torrens to see if they were okay and started turning up. No one told them to. They worked through the night to reboot the power. You don't hear the state government talking about that, or AGL." Back in the day, he said, they had intercoms on every line. "The whole network was connected. Now they have to go to the top of a bureaucracy to get a message down the line. But electricity doesn't work like that." As employees at Torrens Island worked, AEMO instructed Engie to start ramping up at Pelican Point gas plant as well.

By midnight, 80 to 90 per cent of South Australia's power had been restored. But in the morning, as the storm lifted and people surveyed the debris, starting the clean-up, another darkness had fallen, a murkier kind, brimming with words such as baseload, intermittent, synchronous, dispatchable, asynchronous, inertia, peakers.

*

In 1997, Carl Sagan wrote:

> We've arranged a global civilization in which most crucial elements profoundly depend on science and technology. We have also arranged things so that almost no one understands science and technology. This is a prescription for disaster.

To complicate matters further, there are many people who are strongly motivated to exploit these ever-widening gaps in public knowledge.

After the "black system" in South Australia, these gaps in knowledge became apparent – and it didn't take long for the Coalition to take advantage of this.

It started, unexpectedly, with the ABC. Just an hour after South Australia was without power, the chief political correspondent, Chris Uhlmann, was on television and radio surmising that the cause of the blackout was the wind generators.

> Forty per cent of South Australia's power is wind-generated, and that has the problem of being intermittent – and what we understand at the moment is that those turbines aren't turning because the wind is blowing too fast.

It broke the renewables sector's heart. The sector had long been accustomed to attacks from the usual suspects – the Coalition and the Murdoch media – but Uhlmann's report on the ABC was keenly felt as a betrayal. The sector itself had voiced concerns about insufficient upgrading of the grid to support wind and photovoltaic solar. It wasn't only what Uhlmann was saying; it was also his framing of the situation. To many supporters of renewables, it felt as though they were being set up for a rerun of the wind-farm attacks of recent years.

They were right. Straining at the leash was Barnaby Joyce, deputy prime minister. Like many of the emergency services workers in South Australia, he was up bright and early. "[Wind power] wasn't working too well last night, because they had a blackout," he said on ABC radio. Less than an hour later, he was on Sydney radio station 2GB saying the same thing. Turnbull waded in, taking the opportunity to lecture Labor state governments, which had set "extremely aggressive, extremely unrealistic" renewable energy targets. Never mind that in the absence of leadership over the past two decades, 1.5 million Australians had opted to install solar panels on their roofs – including a military-sized operation on the roof of Turnbull's Point Piper mansion in Sydney. Or that Turnbull had recently signed the Paris agreement to reduce emissions in

Australia but implemented no policy to achieve it, basically outsourcing the heavy lifting to the states and their renewable targets. Another glaring omission, later pointed out by Danny Price, co-designer of the national energy market, was that South Australia's apparently "reckless" rollout of wind farms was in fact the doing of the federal government: "All but one wind farm that's gone into South Australia has gone in because of the Commonwealth Government scheme and not because of the State Government."

And on it went. "Energy security," said Josh Frydenberg, Minister for the Environment and Energy, was the federal government's "number one priority." AEMO's preliminary briefing in the early hours of the morning had advised that all that was known was that an "unprecedented" natural event had taken transmission lines out of the grid, but none of this got in the way of the Coalition's attack on renewables.

Scores of people appeared to join Twitter solely to comment on the blackout. "Bring back coal to SA!" tweeted Gemma from Double Bay, "this is bloody unacceptable to have a whole state sitting around in the dark. Aren't we a developed country?"

It seemed to be infectious. Also new to the Twittersphere, Maddison Dunshea ("HR & Recruitment. Calling Sydney home. Adrenaline junkie.") wrote "Senator Xenophon says SA's arrangements are a textbook case of how not to transition to renewable energy. Quite right @Nick_Xenophon."

Or did she? BuzzFeed's Mark Di Stefano smelled a rat. After doing a search on one of the new Twitter user's profile photo, he traced it back to a stock image website. Soon he discovered all the profile photos of the new Twitter users who had joined to comment on the blackout were stock images from the same site.

Then he followed another thread. All the new Twitter profiles followed two accounts: the first a Bangladeshi data-entry specialist who markets his online services – forum posting, content writing – for US$10 an hour. The second was a branch of the Chamber of Minerals and Energy of Western Australia, a mining lobby group.

The new Twitter users had particularly enjoyed and retweeted a clip from *The Bolt Report* on the night of the blackout, in which Andrew Bolt interviewed Alan Moran from Regulation Economics (Institute of Public Affairs), who ventured that the power would have stayed on were it not for renewables. BuzzFeed asked the mining group for comment. It denied it had paid for dozens of fake Twitter users to be created. Within hours of BuzzFeed publishing the story, complete with screenshots, the new arrivals to Twitter vanished.

In the Adelaide carport, the two politicians refused to look at each other.

It was the launch of AGL's federally funded "virtual power plant" and Geoff Perkins, a retired IT director at an Adelaide school, had enthusiastically shown reporters the large storage battery attached to his carport wall. It was hooked up to software that could share the power he collected from the solar panels on his roof with 1000 other houses. AGL's Andy Vesey, state premier Jay Weatherill and his energy minister, Tom Koutsantonis, convivially took turns to answer questions while AGL staff kept looking at their phones. Josh Frydenberg was late.

The Coalition's attack on Weatherill's "ridiculous" renewables target and his disregard for "energy security" had been sustained for almost three weeks by now, a siege conducted with relish. Frydenberg said, "South Australia has been a basket case when it comes to energy policy" and accused the Labor government of imposing third-world conditions on consumers. It all provided good cover for a desperate ten-day rollout of a long-absent federal energy policy that at times seemed made up on the spot.

In turn, Weatherill had announced his own energy policy. "South Australians," he said, "are not prepared to put up with being ridiculed and have the finger pointed at them by a federal government that has abdicated its responsibilities." The South Australian way, Weatherill revealed, would be to put $550 million towards a huge battery farm, to build a brand-new and state-owned gas plant with fixed supply, and to create new ministerial powers to direct networks and the energy market operator.

Finally, Frydenberg's car pulled up in the suburban street and he walked up the driveway. "Hashtag awkward," Koutsantonis said to journalists and photographers as the minister was seated on a little white chair next to Weatherill for the speeches. Then, as the press conference began, the two politicians stood side-by-side, shoulders touching, refusing to look at each other.

REPORTER: Was your [Snowy Mountain] announcement today designed to embarrass the state government?

FRYDENBERG: (*laughing*) Well, to be honest I didn't even know Jay was going to be here, but I welcome his presence because we need to work together, federal and state governments, the private sector and elsewhere, to ensure the stability of the system.

Beside him, Weatherill was tightly coiled, fury just under the surface as he waited his turn to speak. Turning his head, Frydenberg gave the impression of talking to the premier, but still there was no eye contact.

REPORTER: What about the expansion, the Snowy Mountain expansion announcement – was the timing of that designed as one-upmanship on the state government?

FRYDENBERG: (*laughing*) N-no, well, I do point out 2000 megawatts is twenty times the size of the battery there, Jay, but that's a national scheme ... it's got nothing to do with the South Australia announcement.

The old my-megawatts-is-bigger-than-your-megawatts taunt. Over to Weatherill.

REPORTER: Premier, do you find this a bit galling?

Weatherill woke, as if from a dream. It took a second or two for him to ramp up, to channel his adrenalin – but when he did, whatever your opinion on the facts, it was exquisite to watch.

I've got to say, it's a little galling to be standing here next to a man that's been standing up, with his prime minister, bagging South Australia at every step of the way over the last six months. To be standing here, on this occasion, him suggesting that we want to work together. It is a disgrace, the way that your government has treated our state. It is the most anti–South Australian government we have seen from a Commonwealth government in living memory.

Frydenberg stared straight ahead, clamping his mouth shut. "What we have," continued Weatherill, "is a national energy market that is broken."

> We had a prime minister that came in here, to this state, during the course of the last federal election campaign, celebrating our leadership in relation to renewable energy and then taking credit for it through his own renewable energy target, and for you to then turn around within a few short months when there is a blackout and point the finger at South Australia for the fact that our leadership on renewable energy was the cause of that problem is an absolute disgrace.

Bang. For isn't this the problem – we can talk about asynchronous, peakers, synchronous, export gas prices all we like, but the crux of the issue is a Liberal Party that used the image and personal history of Malcolm Turnbull, a centrist, an enthusiast for climate action and innovation, to get back into office, even though this image is entirely contrary to his party's true beliefs and commitments? Turnbull himself has relied on this image of himself to stay in the job, damping it down a little but not too much. At the National Press Club this year, Turnbull told the media he was "stripping the ideology" out of the approach to the national grid – as if the party he is leading is not steeped in an ideological bog.

*

In the *Conversation* recently, Rodney Tiffen wrote about politics in Adelaide in the 1950s, when the premier, Thomas Playford (hence the name of the power station in Port Augusta), and Labor leader Mick O'Halloran faced each other in four elections – and dined once a week to discuss Playford's plans.

> O'Halloran remained Labor leader until he died in 1960. Playford wept openly when told of the death, and was a pallbearer and speaker at O'Halloran's state funeral.

I have heard similar anecdotes across the board, not just about Playford and O'Halloran, but about Australian politicians in general. There was a time, it seems, when the major parties could agree and argue the details. Of course, this is not to say there were no profound disagreements in Australia's political history – there have been many, but simply for the sake of disagreeing? This seems an extreme pathology of modern politics.

In 2007 Rudd had the nation in the palm of his hand when he was swept to power. The public was overwhelmingly in favour of ratifying the Kyoto Protocol, and a survey by the Australian National University showed seven out of ten Australians wanted action on climate change. Not only that, but Rudd had in Malcolm Turnbull an Opposition leader who was keen to negotiate. There were others too – the Greens wanted to put their all into it – but Rudd wouldn't even talk to them. Instead, he squandered the opportunity. Why? It was just too fun to go into parliament every day and drive a wedge between Turnbull and his party. The result? Tony Abbott: a new Opposition leader whose sole mandate was to oppose everything.

Where to begin with Abbott's war on renewables? Perhaps with his and Joe Hockey's casual comments on "visually awful" and "utterly offensive" wind turbines, as if entire valleys of expanding open-cut coalmines displacing "overburden" such as vineyards, olive groves, grain crops and generations of farming families are attractive. Not that wind farms have the same freedoms as did those mines: the Coalition under Abbott ensured that anyone owning land within two kilometres of a proposed wind farm could veto it. In 2015, after Abbott told broadcaster Alan Jones that wind turbines were "visually awful," he was asked if he'd ever visited one.

"Well, I was on Rottnest Island a few years ago, and I cycled around the island most mornings and my path took me almost directly under the big wind turbine which has been on Rottnest Island for some time," he replied.

The Rottnest Island turbine, a lone tower that was installed in 2004 thanks to substantial funding from the Howard government, has saved the island over $3.5 million, which would otherwise have been spent on fossil fuels. It powers a desalination plant that provides 80 per cent of the island's drinking water (the rest is made up by rainwater). Referring to it as a whirring white giant, people on the island gave a spirited defence of the turbine, and the WA tourism office told journalists the prime minister's complaint was the first the turbine had received in its eleven years of labour.

"Can I be a little indulgent?" Joe Hockey had asked Alan Jones on the shock jock's radio show a year earlier. "I drive to Canberra to go to parliament and I must say I find those wind turbines around Lake George to be utterly offensive. I think they're a blight on the landscape."

It was a concerted campaign, cloaked in aesthetics and newfound fears for birdlife, but in truth intended to protect the status quo – aka coal and gas.

Indeed, way back in 2004, the ABC obtained leaked minutes from a meeting that Prime Minister John Howard had with twelve companies to discuss climate change policies and the possibility of developing a fund for low-emission technology. The group handpicked by the Liberal government was known as the "Lower Emissions Technical Advisory Group" – it included Exxon Mobil, Rio Tinto, BHP, as well as fossil-fuel-intensive industry such as Origin Energy, Alcoa, Amcor, Boral, Holden, Energex and Edison Mission.

After the leak to the ABC, the industry minister, Ian Macfarlane, defended the advisory group, saying the government "makes no apology for consulting with industry across the board." Yet some industry consultations are more fruitful than others. Six weeks after the meeting, the government released its *Securing Australia's Energy Future* white paper, which froze the 2 per cent renewable energy target and funding for proven renewable technologies, while handing over $500 million to develop "clean coal" technology. It also reinstated Australia's opposition to the Kyoto Protocol.

Within two years of the Howard government's decision to keep the target at 2 per cent, renewable energy projects, many worth half a billion dollars in investment, were shelved. Under Labor, they picked up again, but the pressure on wind farms merely shifted. A proliferation of groups, with names like the Landscape Guardians, the Western Plains Landscape Guardians, the Spa Country Guardians of Hepburn, the Waubra Foundation, began to pop up, a so-called "grassroots" swell. One group had its own medical director, who spruiked the existence of wind-turbine syndrome. Another took out advertisements warning of this "disease" – in 2010, Adam Morton reported in the *Age* that opponents of a wind farm in Waubra had paid for ads in the weekly *Pyrenees Advocate* warning of an illness that was said to include "sleep disturbance, nausea, headaches and increased heart rate."

> No evidence was given to back the claims, but the advertisements heightened concerns in the local community and were blamed for sparking rumours that Waubra Primary School was to shut due to concerns that turbines several kilometres away were affecting children's health.

Tug on a thread and it was all part of a Russian doll scheme whereby "Guardian" groups were supported by the Australian Environment Foundation, another astroturfing group claiming to be "grassroots," which was in turn set up by right-wing think-tank the Institute for Public Affairs. For the most part the websites were well designed but curiously devoid of people – although listed directors or chairs often had links to the Liberal Party, and in a couple of cases direct links to the mining industry. Curiously, the Landscape Guardians were missing in action when it came to fighting for the rights of people in regions where coalmines were expanding voraciously.

In parliament, Tony Abbott was tabling people's electricity bills and flapping them with mock rage. It was the carbon tax, renewables, the wind – anything but the rort carried out by the traditional networks. In fact, since 2009, electricity networks had been on a spending spree, clocking up $45 billion, blinging up poles and lines, and building new substations and charging it back to consumers at substantially higher prices. In 2014 journalist Jess Hill laid out the bones of the scam in the *Monthly*:

> So how were these networks allowed to blow billions of dollars on infrastructure we don't need? Here's how it worked. Every five

years, the federal energy regulator grants the distribution and transmission network companies an allowance to spend on capital and operating costs. All the networks have to do is produce a spending proposal that looks "reasonable" – it's up to the regulator to prove that it isn't.

According to the Australian Bureau of Statistics, the electricity industry's profits rose by 67 per cent between 2007–08 and 2010–11. In this same period, electricity bills rose 40 per cent. New South Wales and Queensland – states with the lowest penetration of renewables and the highest amount of infrastructure bling – saw the largest increases. Hill continued:

> Thanks to the actions of the electricity industry, a growing number of Australians are struggling to pay for a basic necessity. In every state, disconnection rates are rising – in NSW, for example, the number of disconnections rose by 25% in 2012. The Australian Council of Social Service says some of Australia's 2.2 million low-income earners are struggling to pay their power bills. These consequences were entirely predictable.

People were struggling – but it had nothing to do with the carbon tax. Tony Abbott knew this, but also knew he was on a winner.

On taking government in 2013, Abbott repealed the carbon tax and reduced the renewable energy target. Frankly, he wished the Howard government had never brought it in in the first place, he told Alan Jones. Investment in renewables again ground to a halt. A company in Mildura, in Victoria's far north, had successfully completed its pilot project and was ready to begin a solar project that would have powered 40,000 homes. It was, like so many, abandoned.

In 2014 Abbott rejected a "Christmas present" of solar panels for Kirribilli House, arguing that they would be too costly to clean and posed a "security risk." His personal security was perhaps his highest priority as prime minister, with an armoured, bombproof BMW purchased for

$500,000 to shuttle him around with his chief of staff, Peta Credlin, who was constantly fretting for his safety.

And so, needless to say, it has been a long two decades for renewables. Yes, an entire state blacked out and it is a problem – as Chris Uhlmann pointed out, the last time this happened was in 1964.

Yes, we are not allowed to make jokes about blackout conceptions anymore; yes, parts of South Australia suffered continued blackouts over yet another "hottest summer on record." (So did parts of New South Wales, but the state relies mostly on coal for power, and wind, solar and hydro saved the day.)

Over Christmas in Port Augusta, almost a year's worth of rain fell in an hour and destroyed the resin on the ash dam. Three days later, gale-force winds blanketed the town for a week in a thick cloud of ash. Then another storm rattled through the southern state, power blacking out again in sections as people gasped for a breather from the endless weather events. And Treasurer Scott Morrison strode into the Australian parliament caressing a lump of coal, urging ministers to "touch it."

So when, in October last year, Uhlmann wrote another column on South Australia's energy problems, linking to it on Twitter and typing, "This should keep the pitchfork crowd busy for days. Knock yourself out," he was playing the victim when he knew otherwise. By claiming he was being attacked by rabid renewable energy ideologues, he was ignoring the long and sorry history that led us here.

In 2008 petroleum giant Shell released two scenarios – Blueprints and Scramble – that modelled the world until 2050. For decades the company had been strategising – and as early as 1982, climate change was being factored in.

Scramble is a scenario that has come about from poor regulation, politicians too fearful to take any popularity hits, and countries unwilling to curb harmful traditional means of economic growth. Global cooperation begins to break down as rich nations compete to secure and drain remaining deposits of fossil fuels. Protests and activism give rise to increased security measures. Agricultural groups begin producing biofuels, which compete with food production, and this sees increases in food prices and deforestation. Global tensions are high, climate change is kicking in, blame is being laid, the poor haven't become rich. By 2050, whatever cards the climate deals, responding is bound to be expensive.

The Blueprints scenario starts much the same way. "It is a slow process at first, two steps forward and one step back," Shell wrote in its report.

> There is almost as much political opportunism as rational focus in early developments. Many groups try to circumvent, undermine or exploit the new regulations and incentives for alternative energy paths. In places, uncertain regulatory outlooks discourage developments. But as successful ventures emerge, halting progress develops into a larger and larger take-up of cleaner energy such as wind and solar. As more consumers and investors realise that change is not necessarily painful but can also be attractive, the fear of change is moderated and ever more substantial actions become politically possible. These actions, including taxes and incentives in relation to energy and CO_2 emissions, are taken early on. The result is that although the world of Blueprints has its share

of profound transitions and political turbulence, global economic activity remains vigorous and shifts significantly towards a less energy-intensive path.

Blueprints, Shell wrote, is its preferred scenario – though Shell's vice-president, Jeremy Bentham, later said that Shell would position itself to remain profitable whatever course events took.

On the release of the two scenarios, Jeroen van der Veer, the CEO of Shell, gave a series of interviews. "People always think ... the market will solve all of it," he said. "That of course is nonsense." Journalist McKenzie Funk summarised van der Veer's argument: "Global cap-and-trade agreements were urgently needed. Efficiency standards should be imposed. All this would require government regulation."

Where does Australia stand on these two scenarios? In which direction are we heading? More pointedly: are our governments willing and able to regulate in the public interest?

*

In a radio interview last December, Josh Frydenberg said that the Coalition would consider expert recommendations and look at an emissions intensity scheme (EIS) to ensure that emissions targets were met, electricity remained affordable and the "lights stay[ed] on." An economic tool, an EIS is designed to increase the cost of electricity produced by coal and gas, and to bring down the cost of less-polluting electricity from renewable sources. This is triggered by requiring companies with levels of pollution above a set benchmark to purchase credits from cleaner energy sources. Over time the benchmark is lowered until it is at zero emissions.

There is widespread support for the EIS – among those pushing it are the Business Council of Australia, the National Farmers' Federation, energy networks AGL and Energy Australia, and the CSIRO. Others, however, such as energy economist Bruce Mountain, have grave doubts. Mountain described the EIS as the "third act" in a sorry saga. "What a depressing and tedious play it has turned out to be," he wrote in *Renew Economy*:

> In Act One, an Emissions Trading Scheme took centre stage. An Emission Intensity Scheme scheme had a walk-on part in Act One, but the leading actor in Act One, Senator Wong, called it a mongrel and it was booed off the stage by all and sundry (except Senator Xenophon and later Leader of the Opposition, Malcolm Turnbull). Of Act Two – "scrap the carbon tax" – the less said the better. Now in Act Three, ten years later, the Emission Intensity Scheme has taken centre stage ...

Mountain argued that an EIS could have two unwanted consequences – first that it will merely trigger a transition from coal to gas, and second that it will lead to an excessive and unsustainable boom in renewables that could undermine public support for a clean energy future. Whether these concerns are mechanical and an EIS could be adjusted accordingly, or they are an innate part of the EIS's structure and unsolvable, remains to be seen. However, weighing up the pros and cons is surely part of the process? Not if you're in the Coalition.

In thirty-six hours, after a party revolt led by Christopher Pyne and Cory Bernardi, Frydenberg was back on air saying that an EIS was not on the table after all. He told reporters, "I have a position that is very clear, that we will not be adopting an emissions intensity scheme."

Turnbull told media he did not know why Frydenberg had originally said what he had. This was odd, considering Turnbull often refers to himself as a businessman – not only were the experts recommending an EIS, but so too were major companies and their peak bodies. Turnbull has also proclaimed his practical, pragmatic approach to climate and energy issues. But in this case he feigned ignorance: "You will have to ask Josh Frydenberg that."

This is not to say the government was doing nothing. After the blackout in South Australia, the Coalition announced an energy policy over ten frenzied days to cover up the fact that the federal government hadn't tried to do anything with the electricity grid – except play politics with it. Ministers were dispatched far and wide to issue statements, ping-ponging thought bubbles across the country. It was an emergency, a time for drastic measures, a self-important "We'll fix it," as though it wasn't the inept and inert state of federal politics that had allowed it to break in the first place.

Having let untruths about electricity costs run uncorrected for so long, the prime minister now found himself in an unusual position. As part of the ten-day policy rush, Turnbull was forced to bring the gas industry to heel, lining up its chiefs like schoolboys as he reminded them of their social responsibility.

In fact, there is plenty of gas, but most of it is being sent offshore thanks to the new LNG plants on Curtis Island in the Great Barrier Reef. Once hooked up to the international market, gas companies did what companies do: they saw a profit to be made and jacked up domestic prices to match what they got from exports.

"It's a neat trick, really," Waleed Aly wrote for Fairfax. "Take a country with enough gas to supply itself 'indefinitely,' send the vast majority of it overseas, refuse to sell locally at a fair price, create a domestic shortage, then demand access to some of our most environmentally sensitive resources as though it's an emergency measure."

Malcolm Turnbull noted that two of the suppliers had given a "commitment to being net domestic gas contributors, as part of their social licence." Pop! There goes another free-market shibboleth – that if everyone pursues their own profit, then the profit of all will be maximised.

Turnbull also reminded them, "The Commonwealth government has enormous power in this area, as you know." This, clearly, was "strip[ping] the ideology" from Australia's energy debate.

Earlier in the year, outlining his vision for Australia at the National Press Club, Turnbull mentioned "climate change" only once, briefly referring to the Paris treaty. Presumably this too was part of his non-ideological plan.

The problem is that climate change is both fact *and* ideological. In 2016, on ABC TV's *Q&A*, an exchange between Naomi Klein, the Canadian writer and climate activist, and James Paterson, Liberal senator and former employee of the Institute of Public Affairs, illustrated this complexity. In discussing the IPA's role in disseminating misleading information about global warming, Klein said to Paterson:

> You're a free-market think-tank ... The reason why free-market think-tanks, who are not scientific organisations, are the ones spreading the doubt is because if it is true, your whole worldview collapses. This whole idea of pushing deregulation and privatisation and government-get-out-of-the-way falls apart because we need to manage this decline. We need to make sure that people aren't left behind.

The left cheered. But Paterson's reply was illuminating.

> On the contrary, Naomi, I think it equally applies. If it applies to us, it applies to you. I mean, you have openly said that one of the great things about climate change is it's a great tool for bringing down the capitalist system. And you want to undermine the capitalist system. Your whole recent book is about "this changes everything." What changes everything? This is the opportunity to change the political debate about capitalism and free markets because climate change is a great tool to do that. That is your own view, which you've espoused many times.

Paterson is right. Action on climate change does require a transformation, not just of the electricity grid, but of the minds of the politicians who intend to lead these changes.

Bill Shorten is a case in point. Labor has announced a national 50 per cent renewable energy target by 2030. In a recent speech to the Bloomberg New Energy Finance forum, Shorten mentioned "climate change" eleven times. "Consider the basic question of hardware," he continued:

Right now, three-quarters of Australia's coal and gas-fired generators are operating beyond their design life. Some are over fifty years
old. Not many of us drive fifty-year-old cars or work with fifty-
year-old computers or rely on fifty-year-old medical thinking when
we get sick. I'm sure that none of you have, on your person, a fifty-
year-old phone. We update, we modernise, we plan for the future.

It was a good speech – one I'd venture that Turnbull would like to have
given. But Shorten too insisted this change was merely "practical – not
ideological." And yet without an ideology, without underlying values, his
words are hollow. Hence Labor's bind over Adani's proposed mine.
Shorten has withdrawn support for the federal billion-dollar loan, but not
for the project itself – as if the average punter won't be able to put their
finger on the dissonance that is a political party ready to fight climate
change and happy to open up a new coal seam.

The recent notorious commercial promoting Shorten had the Labor
leader standing next to a group of mostly white Australians (an Asian
woman is hidden in the background). There was a young male apprentice
wearing a hi-vis t-shirt, some older tradies, women with lanyards hanging around their necks, and the slogan "Australians First" for jobs. The
commercial went to air in Queensland – the rest of Australia wasn't meant
to see it. It was a classic two-faced bid for popularity. Turnbull and
Shorten may say there is no ideology in their policies, but this can leave
one wondering, what remains? A desire for power, a painted portrait in
parliament, an entry in the history books? But for what? Just for the sake
of it?

*

Values are one thing. Then there are the vested interests and their powerful, often hidden influence – what Shell called the "many groups [that]
try to circumvent, undermine or exploit the new regulations and incen-
tives for alternative energy paths."

Today Gina Rinehart is also a cattle-station owner. You can see her in a red wide-brimmed hat, boot kicked up on a rusted metal-bar fence. With the purchase in 2016, with Chinese partner Shanghai CRED, of the Kidman cattle empire, she set her sights on the export beef industry. But there's more going on here. Fracking. In 2015 the Country Liberal Party led by Adam Giles in the Northern Territory granted approval for Jacaranda Minerals, owned by Rinehart, to start gas exploration. In the press release, the Country Liberals emphasised their commitment to jobs – especially Indigenous ones – with these "first petroleum exploration permits on Aboriginal land managed by the Northern Land Council." The mines and energy minister, David Tollner, added, "This demonstrates that traditional owners and the resources industry can both win when they work together." But as the *Saturday Paper*'s Mike Seccombe reported, there was a problem: many of the traditional owners claimed no knowledge of the "deal to which they had supposedly agreed to." Immediately, 380 traditional owners lawyered up and the Lock the Gate Alliance moved in.

It truly is annoying.

The following year, in the lead-up to the territory election, about a third of the territory's pastoralists released an open letter urging electors to vote against fracking for natural gas. By a landslide, the Labor Party won the election after promising to ban the practice. Already Rinehart was part of a $2.7 billion lawsuit against the Victorian government for banning unconventional gas drilling. And it was particularly frustrating because she and Giles were so close. Before the election, Giles informed voters that a hospital Rinehart was said to be considering funding would only go ahead if he remained leader.

What to do, what to do?

Within a year, Adam Giles took a job with Rinehart's Hancock Prospecting, joining another former Liberal politician, Sophie Mirabella. The revolving door in Australian politics is often spoken of, a harmonious shifting of bodies in and out of politics, fossil-fuel industry groups, energy and mining companies. Rules that were meant to stop politicians and bureaucrats from jumping into lobbying roles so quickly they are knocking on their own door and giving themselves a handshake are as limp as old elastic. I hesitate to bore you with the details. There's Labor's Martin Ferguson, Gary Gray, Greg Combet and Craig Emerson, the Nationals' Mark Vaile and John Anderson. The former Liberal resources minister under Abbott, Ian Macfarlane, became head of the Queensland Resources Council just months after Abbott spoke of his efforts to repeal Labor's mining tax. "It was a magnificent achievement by the member for Groom in his time as minister reborn, as it were," said Abbott. "I hope this sector will acknowledge and demonstrate their gratitude to him in his years of retirement from this place." They sure did, offering a $500,000 salary to go with his $150,000 parliamentary pension.

However, these former ministers are just the surface of a much deeper nexus – political staffers are highly sought after within the fossil-fuel industry, and vice versa, it seems. In February this year, Turnbull appointed Sid Marris, formerly at the Mineral Council of Australia and a News Corp bureau chief, as an adviser on climate change, energy and resources policy. Bill Shorten's former chief of staff Cameron Milner had previously lobbied for Adani through the "strategic services" company Next Level. Milner is now back at Next Level, with his business partner, David Moore, a former chief of staff for the Queensland LNP and chief of staff during the Howard government for Mal Brough. The two offer a range of lobbying services such as "Shaping legislation through ethical persuasion and advocacy." Many might say this is the way the world works: politicians must be accessible to industry and it's inevitable for relationships to form on common ground. But it only seems to work one way.

Contrast this with the relationship between, say, an environment minister and a scientist or academic. Last year marine scientist J.E.N. "Charlie" Veron, described by David Attenborough as "one of the world's greatest scientific authorities on coral and coral reefs," planned a visit to Canberra to speak to politicians about just how dire the situation is for the Great

Barrier Reef and to lobby for an urgent shift to renewable energy. Environment minister Greg Hunt was unable to meet him, but Veron did speak for an hour with Hunt's adviser. When Veron appeared on the ABC a few weeks later to discuss the bleaching and bemoaned the lack of political action, he received a phone call from Hunt the next day. "He was berating me and asking why I didn't speak to him first before going to the media," Veron told me. "He didn't even know I'd been in Canberra and spent an hour with his adviser."

It was difficult not to gag when in a speech at the Melbourne Mining Club last year Rio Tinto's Jean-Sebastien Jacques called on Australia to "Open its doors. Open its heart. To the mining industry of the future." He went on to chide, "Now come on, I know you Aussies are competitive in cricket, in rugby – why not corporate tax rates?"

*

Some political commentators suggest Australians need to watch what Turnbull is doing, not what he is saying – and it is true that the PM's actions can at times seem to be in direct contrast to what he says. In spite of Turnbull's newfound advocacy for and promotion of "clean" coal, federal funding has to date streamed into hydro, battery storage and renewable projects. He appointed Chief Scientist Alan Finkel to oversee an independent review of Australia's electricity market – a vast improvement on Abbott's appointment of Dick Warburton, a former chairman of Caltex Australia and outspoken climate change sceptic, to review the nation's renewables target.

And in January, when the Australian Energy Market Operator hired a new chief, Audrey Zibelman, Turnbull deliberately looked the other way, perhaps so as not to draw his party's attention to the appointment. In her last job in New York, Zibelman had been tasked with managing a delicate and contentious transformation of the traditional grid into a modern system of microgrids, and ensuring New York was on track to cut emissions by 40 per cent by 2030.

Perhaps it takes a crisis to trigger an opportunity. But still the dissonance that is Turnbull can be confusing. Suddenly, in the middle of the turbid discourse between the Coalition and South Australia, he was in a hard hat standing next to a part of the enormous Snowy Hydro scheme. Introducing Snowy 2.0, a $2 billion expansion that would increase output by 50 per cent and add stability to the grid's flow of electricity, he said, "It's making a great Snowy Hydro scheme even greater."

The announcement was startling. It seemed out of the blue. Policy on the run. But Australia's energy debate needed a vision like this, an idea that was clean and inclusive. And the prime minister captured Australia's attention ... for about a day. Maybe two. Then the focus shifted back to South Australia, to the argy-bargy between Weatherill and Frydenberg in a suburban carport.

What happened? This is surely where Turnbull could shine. In Annabel Crabb's *Stop At Nothing*, she recounts how Turnbull, after losing his father in a plane crash, took over his cattle station in Scone and in his grief immersed himself in the study of water. Crabb quotes Turnbull describing a day spent at the farm when he was twenty-eight years old with his young son.

> And it was pouring with rain and all the contour banks were full of water and the dams were spilling and the whole place was moving and the water was moving everywhere and I took him and – he was just a little boy, you know, with a hat and a little Driza-Bone – and we went right up to the top of that hill there and we just walked, we followed the water all the way and I reckon that was the best lesson you could ever give a boy about hydrology and water and landscape, because he actually saw how it worked. And then we got down to the bottom of that gully, where there is a well. Which is obviously tapping into the groundwater, and we looked down and I could show him the water moving through the groundwater, moving through the well ... I love water.

This is his gig. Snowy 2.0 with the iPhone flourish is the prime minister's moment. It's "Seize the day" stuff. But the pitch that would normally follow such a big idea failed to appear. In 2015, when the Liberal Party dispatched Tony Abbott and placed Turnbull at the helm of the government, there was palpable relief in most quarters. Speaking to the press, Turnbull pledged "a style of leadership that respects the people's intelligence, that explains these complex issues and then sets out the course of action we believe we should take and makes a case for it."

This was Turnbull's moment, and the Liberal Party's too. Not just the Snowy 2.0, but the whole thing – an ailing and dysfunctional grid, a complex issue, something for the "adults" to take responsibility for. But instead of leadership, Australians got politics as usual. Cheap shots, culture-war baiting, bad and good ideas lobbed like hot potatoes and lost in the trash talk of low-grade politics. After the ten-day policy spree, Turnbull's poker face returned and he resumed the grim task of negotiating with the vipers in his nest – not forgetting the Nationals. It makes for strange viewing, if one can bear to watch. Whatever one's thoughts on who the real Turnbull is and whether it even matters, it can at least be agreed that he cuts a lonely figure.

"You can blame Turnbull for this. For walking into a political turkey shoot," wrote Laura Tingle in The Monthly. "Or you can blame his colleagues for failing to allow their new leader to establish a viable prime ministership. Or we can blame ourselves for behaving like a crowd out of Life of Brian looking for the miracle of the juniper bush." Tingle was right. The blame for the degraded and hamstrung quality of modern federal politics can be shared around. The Snowy 2.0 re-emerged briefly in the recent federal budget, with the Coalition proposing to nationalise the hydro scheme, setting off another, albeit brief, stoush between state and federal governments.

Turnbull is so close to action on climate change. Almost everyone, including the electricity sector and smelters, wants an emissions intensity scheme so that they can plan and start investing – almost everyone except the people behind him. And now there's all that water, a big Driza-Bone story of icy rivers and mountains and electricity. He's got the Blueprint. He even has consensus on the other side. But he can't say a word. Because to sell the vision of the Snowy, Turnbull would have to say that coal is over. He would have to say the words "climate change."

On the horizon in Port Augusta a tower has appeared, 115 metres tall. Its tip glows white. You can fix your gaze on it from miles away, steer your car towards it. More than 20,000 mirrors are arrayed at its base, cupping sunlight and beaming it up at the tower. Fanning out across the arid land are dozens of greenhouses, with 750,000 tomato plants climbing upwards, roots dangling in hydroponic pipes. No pesticides are used. Ladybirds are encouraged. Rain is collected from the glass roofs, and salt water from the Spencer Gulf is pumped into a desalination plant. This is Sundrop Farms, all of it powered by the solar tower, and about 15 per cent of Australia's tomatoes come from here. The farm employs 150 people from Port Augusta and nearby Port Pirie, mostly as pickers and pollinators. In 2014 the project secured funding from the Australian government's Clean Energy Finance Corporation but then declined it after finding a private investor. When I asked Gary Rowbottom if he was excited by Sundrop's success, a ripple of dejection went through him. "To be honest," he said, slowly and carefully, "it makes me slightly sad. I'd like a job there."

Since 2012, nine coal-fired power stations nationwide have shut down. The process has been rocky, to say the least.

"The transition away from brown coal is not the thing that we fear," the mayor of the Latrobe Valley, Michael Rossiter, wrote last year in the *Herald Sun*. "What we fear is being abandoned." In the Latrobe Valley this abandonment, many say, began in the 1990s, when the power stations there were privatised by the state Liberal government. Over 7500 direct jobs were lost in the region, and without any real government plan (next to none of the $23 billion earned from the sale was reinvested in the region) towns such as Moe and Morwell have struggled ever since. The sale filled a hole in the state's debt, but as high unemployment took hold accompanied by increased rates of depression, domestic violence and welfare dependence across the valley, you might venture that a dozen more holes were pricked.

"Where's Penny?" was the wretched cry from the valley in 2009 when, after two years, no one had come to talk about what action on climate change meant for them. Kevin Rudd and Senator Penny Wong ignored all pleas to visit the region. The refusal to consult was particularly hurtful considering the image of Rudd in his early days as prime minister, sitting cross-legged on the floor and listening avidly to Australia's "best and brightest" at his 2020 Summit. It was even more bizarre that the Labor Party ignored union calls for a "just transition" for mining and power workers.

Then, in 2014, the Hazelwood coalmine burned for forty-five days. Occurring just half a kilometre from Morwell, a town of 14,000 people, this fire capped off the long-felt sense of abandonment in the valley. The mine's fire preparations were found by an inquiry to be woefully inadequate. A firefighter first on the scene described to journalist Tom Doig how they were given rough "mud maps" of the enormous pit, sketches of an area the size of Melbourne's city centre. There was no back-up emergency power or accessible water. All of French company Engie's firefighting and water resources, the firefighter recalled, were dedicated to the working part of the mine.

In this, Doig wrote in *The Coal Face*, Engie was successful.

> At the very beginning of the disaster, Hazelwood Power Station lost 90 per cent of its electricity production for a 24-hour period. For the next forty-four days, it was business as usual.

In Morwell and the towns downwind of the fire, coal ash blanketed the valley for over a month as the slopes of the huge mine glowed orange and trembled with fire. But in Melbourne, politicians and media alike were strangely silent. An air of disbelief emanated from the Liberal state government when residents complained that they couldn't breathe and felt sick. In Melbourne, much of the public seemed unaware of the extent of the fire, while questions from residents in the valley about long-term health were dead-batted by health authorities. People were told merely to stay inside and if you had trouble breathing, go to hospital. It took nineteen days for an official suggestion to be made that children and people over sixty-five in certain areas should consider relocating.

When coal is burned, the combustion is incomplete and particles are carried into the air with the ash and smoke. This "particulate matter" can be caused by all manner of substances, including dust, pollen, ash and smoke. Thanks to the Latrobe Valley's high rate of asbestos-related disease, particulate matter is a familiar fact of life. It can be breathed in, lodge in the lungs, maybe find a way into the bloodstream. In towns and on the land where people are living in the shadow of coal-fired power stations or coalmines, or alongside railways where open wagons of coal chug past, levels of particulate matter far exceed those in the rest of Australia. The long-term health effects of the 45-day fire may repeat those of asbestos in the valley. As for short-term effects, the inquiry found it was the probable cause of eleven deaths.

Engie is currently fighting attempts to recover the $18-million cost borne by the state of Victoria in putting out the Hazelwood fire. Around the world, thousands of coalmines are on fire, all emitting carbon. Of these fires, the most difficult and, in many cases, impossible to extinguish occur underground. The seam can smoulder for years and flare up out of the dark and trigger blazes in nearby forests. In the United States, a fire in an underground coalmine has been burning since 1962. In 1984 the nearby town of Centralia was relocated after roads started to split, causing extreme levels of carbon monoxide. The fire stretches for fifteen kilometres under the ground and could burn for another 200 years.

This year Victoria's Labor government put out a press release, *You Were Right: Inquiry Vindicates Latrobe Valley Locals*, announcing $51.2 million would be spent on a long-term medical study, expanded health services, testing for coal ash in roofs, air-quality monitoring and clean-up of mining sites. Mining company bonds for rehabilitation would be increased by 50 per cent. There are more than 60,000 abandoned mines and mining sites, such as old shafts and tailings dams, across Australia. In New South Wales, approval has been granted for forty-five massive coal pits to be left after mining finishes. Queensland alone has about 15,000 abandoned mines. Of these, 300 are classified as "mega," "large" or "medium." There are creeks that run a milky sky-blue, while the Condamine River has mysteriously started to bubble like porridge as methane rushes out of riverbed fractures.

In an ABC report, Queensland mines minister Anthony Lynham told Mark Willacy that rehabilitation of a site is not necessarily a desired outcome: "Rehabilitation can extinguish a resource." In the industry, this is called "mothballing" – it is a brazen way of avoiding clean-up costs. Another tactic is to handball a toxic project to someone like Clive Palmer. The estimated clean-up cost of the Yabulu refinery and its toxic dams is $100 million. Critics say this is optimistic. Two creeks nearby have high levels of ammonia: not a surprise. It's just that when business is good, these things don't matter and you can bet that when it has to be cleaned up, it won't be Palmer or the former owner BHP paying for it. The mining industry has plenty of "quick exit" experience and the most recent boom will end up costing Australians billions of dollars, and yet, to date, no other states or territories have followed Victoria's belated lead in raising rehabilitation bonds.

Victoria's premier, Daniel Andrews, also announced a $266-million investment in the Latrobe Valley's economy, to upgrade infrastructure, including the train line, and to help power workers transition to new industries. It had been a long time coming.

In Port Augusta in March this year, through the federal Clean Energy Innovation Fund, the Coalition finally committed to a loan of $110 million towards Repower Port Augusta's solar thermal storage farm. Like a cool breeze, a sigh of relief went through the town's key advocates, though tempered with "I'll believe it when I see it." As for Nancy Joy Baluch, born in Port Augusta, who never stopped blaming the power stations for the death of her husband, led the town for thirty years and would dig out a clipping from the 1990s of three boys, their white cricket uniforms black with dust from a wind that had blown up – she died in 2013, still serving as mayor.

"I think she probably hated me," councillor Lisa Ludsman said to me in a matter-of-fact way. "She was the toughest person I have ever met. At the end, she was riddled with cancer and continued to run meetings, never letting on how much pain she was in. I saw it, just once – she'd shifted in her seat and there were tears in her eyes. I think she hated me because she knew she wouldn't see it, that I'd be able to see the solar tower go up and not her."

*

Across Australia twenty-four coal-fired power stations remain. According to the Climate Institute, they'll have to shut down, one by one, by 2030 if we are to meet the targets the Coalition agreed to in Paris in 2016. Earlier this year, when George Christensen linked the damage from Queensland's most recent cyclone with the need to forge ahead with opening up the Galilee Basin for mining, he declared that local tradies were ready, waiting for a "green light." What Christensen did not say is that in reality there are two green lights. There is a choice.

There is "Blueprints," which would need hydro schemes and interconnectors to be built; maybe another underwater cable to connect to Tasmania's Roaring Forties; solar and solar thermal towers; batteries and wind farms; microgrids; and a vastly different grid to replace the one that got blinged up for nothing. There are problems to be solved, challenges ahead – but there's work. Perhaps even more work.

Yet this green light is abhorrent to a certain group of people. To them, the idea of starting transitions in regions with coal-fired power stations is repugnant. Minister Matt Canavan took offence at a public meeting in regional Queensland when asked about a nearby town's partnership with AusNet to trial a minigrid using solar panels and battery storage. They were living in a "fantasy," he said. He took it personally and echoed Turnbull's call that cutting-edge "ultra-super-critical" coal power plants need to be built, a statement Martin Moore of CS Energy, Queensland's main electricity supplier, dismissed. Moore told the ABC he was "greatly surprised" by the proposal. CS Energy had no intention of building another coal power plant. As for the cutting-edge technology, most power stations now operating were already "super-critical" – "ultra" registers only a minor improvement, with emissions still double those of natural gas. The technology, the emissions and the costs just don't stack up here anymore.

As for overseas, the oft-repeated line is that if we don't dig up and sell our coal, others will – politicians usually point over the horizon at Indonesia. What they don't say is that we are over there too. Between 2009 and 2014, Australia's global bank, the Export Finance and Insurance Corporation, gave $1.4 billion to coal projects around the world. To witness "our coal evangelism in full swing," wrote Guy Pearse in the Monthly, "take a look at Indonesia."

> Not so long ago, the biggest threats to Borneo's rainforests were illegal logging and palm-oil plantations. Today, the timber industry is not just being eclipsed by coalmining, the timber companies are becoming coal companies in what has become a coalminer's picnic. And if you go down to the woods today in Kalimantan, the big surprise is the extent of Australian involvement. Leighton leads the charge with contracts to mine over 50 mtpa [million tones per annum] in Indonesia at eight different locations. Within three years, Kangaroo Resources hopes to be mining 10 mtpa from eight mines of its own; a Straits Resources subsidiary envisages production of 20 mtpa in the same region; and White Energy wants to upgrade 15 mtpa of low-rank coal into briquettes for export. Between Australian-owned mines and contract mining, Australian companies may soon be digging up more than one-third of Indonesia's coal exports.

"The Stone Age did not end for lack of stone, and the Oil Age will end long before the world runs out of oil," Sheik Ahmed Zaki Yamani, former Saudi oil minister, once told his colleagues at the Organization of the Petroleum Exporting Countries. It is a quote that does the rounds among those struggling for action on global warming, lifting morale, heartening for its unexpected source. But Australia seems hell-bent on digging up every last piece of coal and ensuring it is burnt, here, there and every-where, whatever the outcome.

*

In Port Augusta, the power stations transformed the town from a port and railway maintenance hub into the town that powered the state. It was in the '50s and the area was – and still is – strung together with cables, transmission towers, pylons and powerlines, all marching outwards. Now the smokestacks, boilers and conveyor belts wrapped in corrugated tin are being detonated, each walloping to the ground in a great big cloud of black dust. Just like they did with the crop duster belching out green resin, residents come out of their homes, driving to the southern edge of town to watch. For generations, the smokestacks were markers for those driving through, signs that they were finally driving out of the desert or about to head in. In the *Griffith Review*, Michael Dulaney described going to Port Augusta before a demolition and looking at the innards of the two power stations laid out on the ground for auction:

> An auctioneer told me a bright-red fire door – ten feet by twelve feet of tempered steel clad with pounded aluminium – was to be repur-posed as the entrance to someone's "man cave." Whoever had the unenviable job of cataloguing this industrial detritus had alleviated his or her boredom by coming up with sarcastic descriptions for some of the more underwhelming items: "Divorce Pack" (three fridges, a microwave, two heaters and a cabinet); "The Trap!!" (a mysterious steel cage contraption); and "quantity grease tins on wall."

As people wandered around, whispering and running their hands over the "semi-used spools of insulated cable," Dulaney wrote of not being quite able to pinpoint if they were there to "pay our respects or make out like carrion."

*

I stretch my neck back to look at the small dome ceiling. It is pitch-black. I am inside the Adelaide planetarium. Australian artist Lynette Wallworth is sitting a few chairs away; we are about to watch her film *Coral: Rekindling Venus*. Everyone is silent, holding their breath. The dome floods with water and I am submerged in the sea. In the blue, seals swim across the dome, flippers pinned to their sides as they twist downwards, whiskers fanning. There are shoals of fish like thousands of orange lanterns, a pulse of luminous coral polyps, forests of kelp with sunlight shining through – a beat, and then the magnificent spawning, millions of streaks ribboning out of the corals, thousands of tiny beads popping like fireworks. It is an underwater snowstorm. Marine scientists who have studied the annual mass spawning of the Great Barrier Reef often gather around trays of coral they have cultivated in laboratories in shallow waters of the reef watching in amazement as all the corals spawn at the same time, in the trays and in the ocean. You can feel it, a researcher told me, with your fingers in the water: the strange feeling of rain rushing upwards. In the planetarium, I close my eyes and let the images run over me.

For many of the people I have met, like Charlie Veron, this would not be enough. "What is privilege?" Robyn Davidson once wrote:

> Surely it's not only the accumulation of goods, not only the assumption that one will have a comfortable, relatively safe and healthy life cushioned by technological advance. Surely it should also include the possibility of standing on a beach, backed by red cliffs, facing an ocean where you can see whales and their calves thrashing up foam, schools of flying fish shooting across the water like handfuls of flung tinsel and frigate birds plunging into the flat blue ... It is something, as Friedrich Nietzsche said of music, "for the sake of which, it is worthwhile to live on Earth."

There are manta rays flying now, great big birds of the sea, wings undulating across the dome. I'm captured here, in the dark, with this world over me. I cannot busy myself highlighting passages in books, reading documents and then reading them again. My thoughts flit with the manta rays, and you could say fossil fuels gave us this, this time to think. They gave us argo robots that now reside at the bottom of the ocean, swimming up to the surface every now and then to broadcast data to oceanographers. In 1958, Vanguard 1, the fourth satellite ever to be launched, carried six solar panels into space. Fossil fuels gave us this. The little rectangles of silicon carried data about the composition of the atmosphere back to Earth for six years. This year in April, the United Kingdom celebrated its first day powered without coal since the beginning of the Industrial Revolution. Fossil fuels gave us that as well. Coal has given us more than global warming – it has shown us the world anew. It has brought together the world's most brilliant minds, intrepid explorers, and provided a new perspective on time.

In 1989 French and Russian scientists living in underfunded shacks in one of the most isolated parts of Antarctica brought to the surface the Vostok core, a cylinder of ice that stretched back over 400,000 years. "It was a truly heroic feat of technology," Spencer Weart wrote in The History of Global Warming, "wrestling with drills stuck a kilometer down at temperatures so low that a puff of breath fell to the ground in glittering crystals."

The expedition for polar ice-cores began in the late 1950s, when scientists in Greenland worked out how to study molecules of ice to learn the temperature on the day that the snow fell. But the cores held many more clues, discovered French glaciologist Claude Lorius, who led many Antarctic expeditions to drill into the ice. In 1964, sitting in one of the threadbare huts, Lorius was struck by a thought. As journalist Gayathri Vaidyanathan wrote:

> One evening, while drinking a whiskey on rocks chipped off an ice core sample, Lorius watched bubbles get liberated from the ice. These bubbles must contain air from the past, he realized. Like tiny fossils, they contained a sample of atmosphere from the time the snow fell. And the oldest ice would reveal the Earth's climate history from its very beginning.

It took scientists another twenty years to figure out how to capture the tiny air bubbles and analyse them. When they finally cracked it, the carbon dioxide in each bubble could be measured and mapped over hundreds of thousands of years. The Vostok ice-core is modern science's equivalent of putting a man on the moon. When analysed, it confirmed the prevailing science of climate change. Vaidyanathan continued:

> The resulting graph of CO_2 levels and temperature over Earth's history was remarkable. It showed that our planet's temperatures have always changed in lockstep with greenhouse gas levels in the atmosphere. When CO_2 levels have been low, the planet has been cool, and vice versa. And the levels of CO_2 and methane in the atmosphere today are unprecedented in the past 420,000 years.

You could say fossil fuels gave us this. They gave us the means through which *scientific* understanding could grow (indigenous cultures still connected to the land will say this complexity has always been known) and confirm that the planet Earth — a name iterated in variations across cultures, with a common meaning, "ground"— is in its entirety a system, a creature in a way, that lives and breathes and regulates itself accordingly. There is no *one* cause and effect: clever feats such as the "nodding donkey," the pumps used to sip oil out of the earth, cannot be understood in isolation. Consequences are far-reaching, ever-changing and complex. Fossil fuels gave us much of this knowledge — they really did. They gave us a picture of both the past and our future — and revealed that in a geological blink of an eye, time is running out. The people, the forces that are delaying what science is telling us and pushing the fossil-fuel agenda are not true appreciators of coal. They are ignoring everything this "magic" black rock and all its variations — gas, oil — has taught us and is still teaching us. Coal led us out of darkness, and now these people want to take us back.

BLACKOUT

> When you're out,
> When you're about
> with everything you bri-ing –
> Do the Right Thing!
>
> Change your way
> Have a better day (c'mon!)
> Everybody si-ing –
>
> Do the Right Thing,
> Do the Right Thing!
>   – Australian TV public service announcement, 1980

There has been a lot of talk about Australian values over the past two decades. The word "unAustralian" entered our vocabulary via predictable and political culture-war baiting that has now become toxic. But what is it, to be Australian? Is it a standard, as in those public service announcements of the 1980s with Australians picking up litter, one man wielding the tip of his umbrella to spear rubbish and putting it in the bin? Or is being Australian an entitlement? The kid in the 1980 "Do the Right Thing" ad shaking his head at a man who threw his betting form on the ground outside a TAB would be liable to get his face smashed in today.

If we were to look to the creche on the hill for guidance on Australian values, it might go something like this: "Don't fucking tell me what I can and can't do, but by god, no way are poofters going to get married on my watch." It's confusing, to say the least. There is a huge hole at the heart of Australian politics and no obligatory Anzac Day speech is going to fill it.

How has this happened? In part it is because Australia has a political system captured by the fossil-fuel industry. It is a political Stockholm syndrome built on donations, royalties, taxes and threats (Rudd's fate still looms large). The vast pensions given to politicians on their retirement were premised on the idea that being a civic representative, making the hard decisions, could lead one to become professionally untouchable. Instead, parliament has become a transit lounge for politicians and their staffers on the way to fossil-fuel companies and their lobby groups. Inertia is the result.

What to do? Ban political donations. Of course we should, although some politicians will get creative. Or, if that is too radical, require all politicians to wear the logos of their sponsors on their jackets, like Grand Prix drivers do. This would include the investments of donors such as Soul Pattinson, a pharmacy conglomerate with vast coalmining interests. This may sound drastic, insanely idealistic, but why? Don't Australians have a right to know where their politicians' values are coming from?

In 2012, BHP's chief executive officer, Marius Kloppers, said federal and state governments had been "absolutely fantastic" in backing the Olympic Dam project. It would be good to know how "fantastic" BHP has been in return. Duck for cover as the howls go up of being anti-industry! But if that is anti-industry, then transparency is anti-industry. If so, then maybe we really should call the assessments process what it is – what the fossil-fuel industry already calls it in a semi-slip of the tongue: the "approvals process."

Here's another public service announcement from the 1980s, part of the "Life. Be in it." campaign: "This man is walking. It's not very hard. Just watch him for a while and you'll get the idea." After showing several activities involving walking, the advertisement continues: "Best of all, walking is free! Try it today. Here is a simple walk to start with. Stand up, walk over here and turn off the telly."

Imagine. Today's politicians can't even ban junk-food advertising between children's television shows, let alone suggest turning it off. Put accurate health labelling on food, list palm oil in the ingredients – in spite of huge public support for such measures, politicians today won't touch them. In Victoria the Liberal state government forced Zoos Victoria to withdraw its campaign to promote recycled toilet paper because it affected industry, while in the Northern Territory Coca-Cola managed to dismantle a recycling scheme. Most Australians are keenly aware that the only time legislation is amended and is hurried through parliament is when it has become an obstacle to the "approvals process." There are piecemeal, fig-leaf gestures, such as the $1 billion given over ten years – well short of at least $8 billion that experts say is needed – to improve the quality of farm water flowing into the Great Barrier Reef. These are not measures to "save the reef"; they are measures to "save face" when people start asking, *what did you do?*

Duty of care? Responsibility? Making the tough decisions? Transparency? *Turn off the telly?* Don't be silly.

*

Does it matter what Australia does? If the planet really is going to hell – if there really are thousands of computer nerds in the United States desperately backing up scientific research and data since it became clear President Trump intends to "burn the books" by deleting swathes of climate research at organisations such as the Environment Protection Agency – if Trump's secretary of state is the former chief of Exxon Mobil, which, by the way, despite earning $3 billion in Australia last financial year, paid no tax – does it really matter what we do? Why not suck up as much as we can from the ground, deplete the watertable and drain the aquifers to sink the land just that little bit further to meet the incoming tide? But if we do choose this, then we had better start taxing the fossil-fuel industry accordingly, so that we can afford the contingency plan. And herein lies a problem. Raise the taxes, charge the true cost of fossil fuels and suddenly they may not seem quite so appealing to mining companies. Coal is cheap because the cost of emissions is not factored in, and because when it comes to expenses such as infrastructure, exploration and rehabilitation, Australians pick up the tab.

But the question remains: does it matter what Australia does? Perhaps the answer depends on what year you pick. In 2002 it appeared Australia had an important role to play in the great global – and moral – crisis that was Iraq's weapons of mass destruction. However, in 2010, when it came to our mining sector contributing to the possible making of WMD, the Labor government signed off on a Howard–Putin deal allowing Australian uranium to be processed in Russian facilities not open to inspection by the International Atomic Energy Agency – despite Russia stating that its nuclear arsenal "remains one of the top priorities of Russian Federation policy."

And in 2017, Australia plays down its role in cutting global carbon emissions. On the home front, Australia's emissions per person are among the highest in the world. Worse, it is possible these are being under-reported. At the Bayswater coal-fired power station in the Hunter Valley, pollution has been monitored from only one of its four generation units. Recently, workers, including a former engineer, claimed they had been instructed to supply lower sulphur coal (less emitting) to the monitored unit while dirtier coal was burnt in the other three. The state Environmental Protection Agency is now investigating all NSW power stations, though whether this watchdog has any teeth is a matter for the state Liberal Party.

As for our international role, various politicians insist that moving to cap our coal exports will not make an iota of difference. If we don't dig the coal up, they say, other countries will. It is true that many coal exporters will continue to supply the market whether Australia is in the game or not. Similarly, countries will continue to dig up their own coal for domestic use. Does this mean Australia is powerless to commit to genuine action on climate change? Of course not. If Australia opens up the Galilee Basin, we will add substantially to the over-supply of thermal coal, causing the price of coal to drop and delaying the transition to clean energy.[2] If we decide to

---

2   Yes, we can throw coking coal into the debate. About a third of the world's steel production uses coking coal – of which Queensland currently provides half. Much of it is going to China as it builds new infrastructure; the other two-thirds of the world's steel is recycled. Future forecasts for coking coal suggest supply can be met by existing coalmines, which has not stopped mining giants such as Shenhua pushing ahead with plans to open new mines, often on farmland, rather than purchase existing ones that can produce the same amount.

leave the coal in the ground, open no more new coalmines and curb expansions, we will help tip the balance significantly towards change. Ironically, we would even protect the price of our current coal exports.

*

American writer Chris Goodall calls it the "switch," the global transformation in how we use electricity. The "switch" is swift or slow, depending on your viewpoint. Up close, the past two decades have felt glacial, a despairing slouch towards nihilism. Take a step back and the transformation is giddying. South Australia has almost reached 50 per cent renewables, and one of the world's largest solar and battery farms in the state's northeast will join the grid next year. The ACT is on target to reach 100 per cent renewable power by 2020. May we all pray for a blackout during Question Time. Households in the ACT also pay the lowest electricity prices in Australia. A local start-up, Reposit Power, has developed software that allows owners of solar panels and storage batteries to buy and sell power, and has just signed a partnership with Tesla. Over 1.5 million houses across the country have solar panels on their roofs, while towns and shires have set targets, some setting up localised grids. It is happening all around the world – slow and swift.

In 2015, in many of China's major cities, people could barely see. The levels of air pollution were so high that they struggled as if in a snowstorm, face masks pulled tight. The thick mass spilled over into neighbouring countries and officials in Beijing issued a red alert later in the year, closing schools, factories, building sites, businesses, due to hazardous levels of particulate matter. At the peak of Australia's recent resources boom, about 1.2 million people died in China as a result of air pollution. Sensing unrest, the Chinese government announced a groundbreaking shift and committed to reach peak emissions no later than 2030, and the city of Beijing pledged to end its use of coal by 2020. With gusto, China rolled out in a single quarter solar capacity to match countries that had been working towards similar goals for years. Prices around the world tumbled. When China moves, the world moves. In the *Financial Times*, Nick Butler, a former head of strategy for BP, wrote, "advances along with falling costs promise to make solar the power source of choice in the 21st century."

But Australian politics is in a blackout of its own making.

In May, Acting Prime Minister Barnaby Joyce, who may as well be Gina Rinehart in drag, and Adani's Australian CEO, Jeyakumar Janakaraj, met in South Australia to make an announcement. Adani would throw the struggling steelworks in Whyalla a lifeline, signing an agreement with Arrium for $70 million worth of steel to build the railway from port to mine.

"Though it would have clearly been cheaper to source the rail from overseas," said Janakaraj, "Adani values supporting Australian businesses and Australian jobs."

Presiding over the deal, Joyce was delighted. He was in enemy territory with a sweet chaser. The Whyalla steelworks is buckling under $2.8 billion in debt. In 2015 the Coalition agreed to lend $50 million to upgrade its machinery, but refused to match federal Labor's pledge of $100 million. The South Australian government put in $50 million.

Now if you oppose Adani's coalmine, you oppose Whyalla.

There are so many cascading angles here, starting with the Coalition slashing $500 million in funding to bail out the automotive industry in 2013, and Treasurer Joe Hockey taunting car companies to leave and seeing nearly 24,000 jobs lost in South Australia. Or that $70 million will only go so far, with the recent mining boom being another nail in the coffin for manufacturing in Australia. "Our estimates show that the mining investment boom is so big," wrote analyst Paul Bloxham of HSBC back in 2011, "that in order to make way, other parts of the economy need to shrink to their smallest share of the economy ever." But the most overwhelming angle is just how far today's politicians are willing to go to create rifts so toxic, so divisive, that Australians are pitted against one another, all so that they can push through a murky project that will open up a coal seam and see a tenth of the world's carbon budget go up in smoke.

*

In Townsville, I have an interview with Russell Reichelt, the chairman of the Great Barrier Reef Marine Park Authority. By a turn of events, I end up in his office and he's in Canberra. We talk via video hook-up. Reichelt picks up his mug of coffee and offers to get me one. He is in Canberra, he says, to brief his fifth environment minister, this time Josh Frydenberg, about temperatures on the reef. He does this every few months. Reichelt tells me about the reef – it is critical, he says, and in decline. "It would take ten to twenty years for it to recover from the bleaching last year," he tells me.

"We've had a decade of phenomenally big storms." Category 5 storms, he says, were unheard of. "There's been two decades of major outbreaks of crown-of-thorns starfish ... we see the reef as being in a long-term critical phase ... It is in decline and it will get worse." Reichelt has a way of talking that is making me sleepy. As if sensing this, he says, "People say, why aren't you upset, you look so calm on television, and of course I'm upset, but I feel like I've been saying this is coming for ten, fifteen years."

Reichelt came under heavy fire in 2014 when the Authority approved Abbot Point port's permit to dispose of 3 million cubic metres of dredged seafloor in the marine park during its Galilee-related expansion. Plumes of sediment can drift over coral and block the sunlight, killing reefs. Two months before the permit was granted, the esteemed water-quality scientist Jon Brodie, who works closely with the Authority, wrote in the *Conversation* that he estimated about 140 million tonnes of dredged material would be dumped in the Great Barrier Reef over the next decade from port development. Brodie leads the Reef Rescue program, working with farmers and graziers to improve the quality of water flowing into the reef. After $200 million in funding, "even more from farmers' pockets," and decades of work, efforts were finally paying off. The program had managed to stop 360,000 tonnes of sediment and agricultural waste run-off. "Yet all that effort to protect the reef," he wrote, "looks like it is about to be swamped."

The Authority's decision to approve the Abbot Point permit sent ripples of shock and anger through the marine science community. There were clear tensions as senior scientists advising the Authority told journalists they had flatly rejected the approval. There was an exit of staff. One director said he resigned in disgust after two decades of service. A conservation group filed a challenge in court. The troops were rallied. At the time Reichelt argued that it was not a new practice: the Authority has always permitted tonnes of dredged material to be left in the marine park. This did not go down well. The then federal environment minister, Greg Hunt, tried to hose down the furore, announcing a ban on dumping dredged spoil from "capital" works in the Great Barrier Reef — brief pause — but maintenance dredging was still allowed. It didn't work: everyone from local conservationist groups and tourism operators through to organisations such as GetUp!, Greenpeace, the Australian Marine Conservation Society and Seed Mob — an Indigenous climate group — saw the bigger picture: it was Coal versus Coral. Everyone except Reichelt, it seems.

I ask him about the approval, about the dredging. "What people forget," he says, "is that five years earlier, not when I was in the job, the Authority had approved nine million cubic metres in one year." By comparison, Abbot Point's three million "did not seem like much." He keeps talking and it's important what he is saying, "We need to pull every other lever to boost the resilience of the reef," and it's good, it really is, the Authority is working on water quality, trying to ensure fishing boats and trawlers obey regulations, "we know they are being breached because the ecosystem has diminished and the remaining fish are smaller," the Authority is trying to implement GPS on fishing vessels to monitor when no-fishing zones are entered, the crown-of-thorns starfish culls, education – which I think is really good, but I'm still sleepy. "Dredging?" I say. An end to dumping dredged spoil in the marine park is down the list of priorities, Reichelt says. What about "every lever," I think, but there's only so much one can do, funding stretched, why am I so sleepy?

I finish up my interview with Reichelt. Nice guy, I think. Knows a lot. In his time as chairman, Reichelt has overseen a five-yearly report that is considered one of the best assessments of the reef's health. In the summary at the front of the 2014 Outlook report, Reichelt wrote:

> Even with recent management initiatives to reduce threats and improve resilience, the overall outlook for the Great Barrier Reef is poor, has worsened since 2009 and is expected to further deteriorate in the future.

He is not hiding anything. It is this report that informs the Coalition of the state of the reef. And yet in 2015, federal Coalition and Queensland Labor politicians successfully lobbied to keep the Great Barrier Reef off the UNESCO "in danger" list, spending $400,000 on travel costs as Hunt and an envoy went to various countries connected to UNESCO to rally (or negotiate?) support. They also demanded that sections concerning the Great Barrier Reef be censored from a report, "World Heritage and Tourism in a Changing Climate." Seeing the final report, one of the authors decided to publish the uncensored draft online, forcing UNESCO to add a sentence to its press release: "At the request of the government of Australia, references to Australian sites were removed from the report."

I re-read the Authority's report – it was good. Thorough. I check again at the front – yep, the Great Barrier Reef is definitely still dying. I realise what was making me sleepy. Reichelt has been pragmatised. He briefs the minister, he briefs the Senate, and he's sleepwalking. It's not fair, but I want to zap him. Wake up! You have to! You have to fight! You are in charge of one of the most beautiful underwater creations in the world, a vast turquoise structure made up of living corals, a section of the earth that is essentially gills, and it can't breathe.

When I was a kid, my mum sat up for long nights with me as I tried to breathe. Asthma, it tightens everything, it feels like being strangled, like a great weight being rolled on top of you, you have the tiniest pinhole through which to draw breath, and all night I would focus on that pinhole, trying to suck up air, tired, so tired. They were long nights, my mum saying, "Breathe, breathe." Then, finally, the sun would rise, warm me up, the weight would release me and I could breathe.

The reef isn't dead. But it can't breathe.

26 May 2017

THE WHITE QUEEN

Anne Aly

When David Marr called to tell me he was writing a Quarterly Essay on One Nation and the politics of race, I was excited for two reasons. First, because I've always deeply admired Marr's capacity to capture in lyrical prose even the most mundane details of Australian politics. Second, because – like many of my parliamentary colleagues – I am uneasy about the appeal of Pauline Hanson and One Nation for some Australian voters. But I'm not prepared to ascribe that appeal solely to the worldwide populist trend captured by the terms "Trumpism" and "Brexit." In his astute analysis, Marr isn't willing to do this either. Like him, I'm more inclined to ask questions about our Australian character and what makes us as a nation vulnerable to the politics of race.

Several years ago I visited Pakistan as a scholar of counter-terrorism and counter-radicalisation. I participated in a roundtable discussion with politicians, activists, journalists and academics about Pakistan's struggle to contain its domestic brand of terrorism and its increasing intolerance of religious diversity. In other countries, I had already heard about the growing Wahhabi influence coming out of Saudi Arabia, and I listened as several people at the roundtable lamented that influence. But then one participant, a journalist, asked a question I had not heard before: "We can talk all we like about the influence of outside forces and their impact. But we have to ask, what is it about us? What makes us so vulnerable?" Indeed. Australians might ask the same question. What makes Australia so vulnerable to that "something grubby," as Marr puts it: the divisive rhetoric and dog-whistling that not only accommodates the politics of race, but also gives it a tacit nod of approval?

When the federal industry minister, Arthur Sinodinos, was questioned about the preference deal between One Nation and the WA Liberals at the March 2017 state election, he argued that One Nation was a new party, one that had "evolved" and was "a lot more sophisticated." Like so many, I was sceptical of this claim. In

QE 66 2017 **117**

a sense, all that's changed is that the Muslim "other" has now replaced the Asian "other" as Hanson's target – "People [who] don't look right." But that is not the whole story. What has evolved is not One Nation or Pauline Hanson, but our response to them. Politics and the media respond differently now than when Pauline Hanson first appeared on the national scene. Politicians especially are much less likely to call out her racism – or even to use the word "racist" when referring to her or her voters.

In his essay, Marr attributes Hanson's support in part to nostalgia. We discussed this when David rang me, as I'm not so sure I completely agree. Nostalgia by definition involves a wistful longing for something that once existed. Did white Australia ever really exist other than in our imagination?

I have never known a white Australia. On the streets of Sydney's western suburbs where I grew up, Australia was inherently multi-ethnic, multicultural and multi-religious. If a white Australia has ever existed, it was only in the minds of the community of white settlers searching for cultural anchorage in a strange and often unforgiving land.

Anxiety about our porous borders dates back to the early days of settlement, when Australia was imagined to be vulnerable to invasion by the "Oriental races" of an overpopulated Asia, who were poised to flood in and take over our land.

When white settlers first set foot here, they had to define their relationship with an unfamiliar and unpredictable landscape. The tyranny of distance from their closest cultural kin in the United Kingdom meant they also had to define their identity as a white satellite in a predominantly Asian region.

The uncertainty of the landscape in which white settlers found themselves – tropical and desert, wet and dry – when coupled with its vast empty expanses, underscored settler fears about whether they could create a culturally and racially homogenous future. Their fraught relationship with the landscape also gave rise to a lexicon of tidal images to depict invasion. Since the nineteenth century, the "other" in Australia has been described variously as a *peril*, a *menace*, an *evil*, a *wave*, a *tide* or an *influx* ready to *invade, inundate, swamp* or *flood* Australia and *annihilate, oppress, obliterate* or *penetrate* the invisible rabbit-proof fences of racial and cultural homogeneity.

Hanson's imagery of "waves" of immigrants swamping Australia's shores and inundating our land thus expresses a persistent anxiety. This anxiety is not so much about the infiltration of physical borders, but the erasure of imagined cultural boundaries by visibly different others – "People [who] don't look right" – whose very presence in Australia threatens to undermine the racial and cultural purity that was once the vision of White Australia.

Her secret has been to tap into the regenerative capacity of fear about our borders: to insert into contemporary Australia the imaginary old, better way of life where Australians were unified by race and by culture – one nation.

Hanson's political success raises some inevitable questions about who we are and how we see ourselves today. Do we live in a new social order, where blatant bigotry, racism and aggression towards "others" are not only considered acceptable, but even sanctioned by social, economic and political structures? No. I don't think Australia or Australians in general are racist. But I do think the insecurity that continues to plague our concept of who we are – the legacy of a persistent cultural anxiety – is like a sleeping but volatile volcano that occasionally threatens to erupt. Hanson provides the occasion.

Anne Aly

Dennis Glover

"*Islam is a disease; we need to vaccinate ourselves against that.*"
— Pauline Hanson, March 2017

"*The Jew is a global plague.*"
— Adolf Hitler, Mein Kampf

There, I've mentioned him in the same breath as Pauline Hanson: Adolf Hitler. As much as people will inevitably complain about the parallel (and clearly the strangely inarticulate One Nation leader is unlikely to get any further than emulating Hitler's evil rhetoric), Ms Hanson rather brings this upon herself. Indeed, given the ugliness of her stupid but obviously calculated remark – which came too late for inclusion in David Marr's excellent Quarterly Essay – it would be gutless of any writer to pass over the comparison.

The very idea of a populist politician describing a vulnerable religious minority as a virus or bacillus must surely ring alarm bells. Anyone who reads Volker Ullrich's extraordinary new 800-page biography of Hitler will be left in no doubt as to what the essence of that alarum is: the politics of the 1930s, over which Hitler triumphed by exploiting racism, is in danger of returning – unless we fight it with courage and intelligence.

This warning to us from Europe has been getting louder for some time. Five years ago, the satirical German novelist Timur Vermes shocked Germany with his bestselling *Look Who's Back*. The premise was simple but brilliant – a petrol-soaked Adolf Hitler wakes up in Berlin in 2011 to find a strangely familiar world: armies of the unemployed, silent anger among the people, a religious minority to blame and a dissatisfaction with the prevailing circumstances that reminds him of 1930. Add the modern phenomenon of political apathy, this reborn Hitler says, and "conditions were absolutely perfect for me." The reawakened Hitler is taken as an impersonator of uncanny physical resemblance and method-acting genius by a shallow media industry obsessed with celebrity; his racialist and anti-Bolshevist tirades are mistaken by the left and the right alike for subtle irony and go viral on "U-Tube"; he is given his own cable television talk-show, filmed in a purpose-built studio modelled after the Wolf's Lair; he is offered a lucrative book deal by a leading publisher; and eventually all the mainstream political parties approach him for endorsement. He rejects them, of course – they're all washed up – and decides instead to start his own party, complete with retro 1930s merchandising that includes the slogan: "It wasn't all bad." He feels he's been given a second chance. History, as Marx famously said, always repeats: the first time as tragedy, the second time as farce, and no Australian could read Vermes's novel without thinking about *Dancing with the Stars* or Channel 7's *Sunrise* and seeing just a little bit of their own country in 2017.

The strength of Marr's *The White Queen: One Nation and the Politics of Race* is those last five words of its subtitle. No matter how much political tacticians and strategists may want to ignore the centrality of race to One Nation, Marr won't let them. He deftly allows them to skewer themselves with their admission that while they know Hanson's appeal is racist, they're only going to talk about economic insecurity. Here, for instance, is a Labor spokesperson, quoted by Marr:

> We can only address this through dealing with their economic insecurities … If you say to someone, "Vote for us because that woman is a racist," we'd be classified as elites. We'll get killed electorally. If all we do is try to address the cultural issues, we'll lose.

While one can appreciate the tactical nature of this line of thought (and there is a need for tough political tactics, obviously), it does involve a degree of intellectual dishonesty and moral evasion that betrays political weakness. Marr responds: "Of course, the big parties could try doing both: confront the racism and deal with the economic issues. But that isn't happening." And this seems to me the essential issue: why aren't the major parties showing a little more directness and courage in combatting Hanson's racism? Can it only be a lack of political ability? After all, in answer to Marr's plea, Gough Whitlam, Bob Hawke and Paul Keating would have managed to walk and chew gum at the same time.

I suspect part of the reason lies in a superficially appealing but inadequate idea that has conquered the political left in recent times: the notion that it can outplay the right with a populist message of its own built around exploiting economic grievances. Left circles at present are abuzz with plans to create a pro-gressive populist project that mimics Bernie Sanders' (failed) primary campaign – one that is both economically egalitarian and socially liberal. Part of me wants to agree – every left-winger wants to believe that all grievances are ultimately economic: take away economic misery and you take away the impulses to intolerance – but deeper down something says no.

It isn't just because the left will likely get it wrong. Trapped as they are in their academic and managerial mindsets, one can imagine the sort of slogan these progressive populists would produce – something like: "For a Gini coefficient closer to zero than to one." Revolts from below generally don't originate in sem-inar rooms. The problem is that this sort of thing misunderstands the nature of populism: populism isn't nice or about promoting mutual tolerance and respect; indeed, "progressive populism" (taken as socially liberal as well as economically egalitarian) seems a contradiction in terms. Populism is about unleashing pas-sions and hatred and blame. As the left discovered in the 1920s and 1930s, you can't out-populist people like Hitler: they have the better formula for stoking resentment and will beat you every time. If there is a lesson from the 1930s – and Volker Ullrich is brilliant on this – it's that you have to be strong in standing up to people like Hanson: never give them an even break or a second chance, never think you can do deals with them or control them or co-opt part of their program, never think that they are getting "more sophisticated" and it is there-fore safe to deal with them as equals, as Arthur Sinodinos recently did. In other words, you need to show moral leadership in dealing with the likes of Hanson, and if you don't they will exploit every weakness to steal your votes. They're not here to play a supporting role; they're here to replace you.

The left's real Hanson problem is hinted at in another piece of evidence in Marr's essay, provided by social researcher Rebecca Huntley:

> The general conversation from the community is that politicians seem like a kind of a club: they all know each other, they all went to university. They see them as highly educated, highly connected, an elite they have never been part of.

Here's Labor's problem neatly stated: it's not a lack of a populist economic message – which its Stiglitz-reading young economists seldom cease talking about – but a lack of connection with the working class and less-educated lower-middle class to whom Hanson appeals most strongly. Because its connection with such voters is increasingly tenuous, and because its approach to economics is largely theoretical, Labor can't really understand and act on the nostalgia Hanson's voters have for the world that the Hawke–Keating years of economic reform helped sweep away. No one can invent a time machine to take us back to that era, but Labor can't seem to project that it understands what might have been lost and why for such people the past has an enduring appeal. The party doesn't seem to understand the basic propositions that when things change they don't always improve for everyone, that the losers can't be waved off as unfortunate collateral damage, and that rising GDP on its own can't fill the holes opened up in people's lives by decades of accelerating economic change. In other words, it doesn't understand the fundamental human appeal of nostalgia: that not everything about the past was bad – a lot, yes, but not everything. One suspects that until Labor can re-establish some psychological connection with its traditional working-class and lower-middle-class base, it will keep losing these voters to the likes of Hanson. Marr is right that we can't bring back the old world of tariffs and White Australia and institutionalised sexual inequalities, and nor should we – but we can try to find new ways of expressing the essence of what made the past different to today: the idea that within living memory Australia offered something for everyone, not just the young and the winners. "Progressive populism" seems to me nothing more than a cynical tactical shortcut.

One thing's for sure: Bernie Sanders' playbook and Joseph Stiglitz's economics won't provide all the answers. Like everything difficult in politics, seeing off right-wing populist challenges like Hanson's takes a moral effort to put aside the factionalism and the infighting and the divvying up of prizes to get back to the business of listening to the people and showing strong moral leadership. After all, as Marr's essay tells us, there's a big vacuum there waiting to be filled.

Dennis Glover

Lachlan Harris

David Marr is at his best when he is delving deeply into the complex personalities of our political leaders, and charting the mysterious schematics of the movements they build. *The White Queen* is no exception to this. It's tough, insightful, meticulously researched and beautifully written. I have seen at first hand just how influential David's profile pieces can be both on his subjects and on the political zeitgeist in which they operate. I have no doubt *The White Queen* will continue this tradition and influence both the way we view Pauline Hanson and the way Pauline views herself. It matters not whether she has read the essay, because it's not reading a Marr essay that changes you; it's the realisation that in being the subject of such an essay you have, in fact, been accurately read.

My response to *The White Queen*, therefore, is not to take issue with what has been written – it's an excellent profile of an important figure – but rather to provide some context with which to read the essay. This context is an understanding that the structural shift in voting patterns towards the minor parties since at least the early 1990s is not simply a Pauline Hanson or One Nation phenomenon. One Nation and its seemingly indestructible leader are an important part of the shift, but they are not the only reason for a tectonic political change.

In the most recent federal election, 3,145,200 people cast their primary vote in the lower house for a minor-party or independent candidate. Only 175,020 of those votes were for One Nation. To put that modest result in perspective, the Nick Xenophon Team won 230,333 votes in eighteen lower-house seats (One Nation contested fifteen seats), the Christian Democrats won 178,026 votes in fifty-five seats, and 380,712 voters cast their primary vote for an independent candidate in the seventy-two seats where at least one independent was on the ballot.

In other words, even if Pauline had never graduated from frying flathead fillets at Marsden's Seafood, the minor-party vote in the 2016 federal election would have been extremely high. Indeed, a close examination of the lower-house voting data over the last thirty-odd years reveals that the shift away from the major parties always survives the implosions to which all minor parties are subject, especially One Nation. The minor-party vote peaks as a particularly potent party rises, and falls after each significant implosion. But every time the peaks are higher, and the troughs not quite so low. This suggests that even if One Nation implodes yet again – a distinct possibility – or if its vote plateaus because of the stunted nature of its political offering, the shift to the minor parties will continue, unrestrained by the political passing, or peaking, of the White Queen.

Why does understanding the structural integrity of the minor-party vote matter for David's essay? Because it will ensure readers avoid conflating the limited political potential of One Nation with the much more durable, and in some ways more ominous, broader appeal of minor parties. David is right to be sceptical about the capacity of One Nation to grow from a protest party to a "programmatic party that's got a stable set of policy issues." The racist, protectionist, exclusionary political philosophy that sits at the core of the One Nation fantasy is perfectly designed to generate headlines, clicks, likes and shares, but it is highly unlikely to scale beyond the minor-party universe. It is too intellectually dishonest to become "programmatic," and Australia has too many decent, middle-of-the-road voters for the One Nation primary vote to ascend to the major leagues. But none of this means that the kind of populist political turmoil we have seen around the world – Brexit, Trump, the elimination of both establishment parties in the first round of the recent French presidential election – will not, or cannot, happen in Australia.

Why? Because the broad-based, enduring nature of the minor-party vote, rather than the measly, capricious and unstable vote for any particular minor party, suggests that it is not the rusted-on One Nation voters who will drive the next big unbundling of Australian politics. Rather, it is the voters at the periphery of One Nation (and on the periphery of all the other forty-five minor parties) who will be the foot soldiers of political change. By weight of numbers alone, this groundswell of voters – who are sceptical, even dubious, about the political legitimacy of the minor parties they are voting for – is the real force to be reckoned with (or resource to be tapped, depending on your political inclination). These reluctant protest voters, driven to the minor parties by chronic dissatisfaction with the majors, could up-end Australian politics and could well be corralled into a fully scaled-up, savagely disruptive, programmatic political movement.

I also agree with Marr's conclusion that the "standard explanation – that these are people left behind by globalisation" does not accurately describe the "decisive component of the Hanson vote." Reduced to its philosophical core, One

Nation is a party set up to take advantage of a xenophobic and racist current that has flowed through Australian politics since the first ballots were cast in our first colonial elections. As George Megalogenis has expertly charted in his book *Australia's Second Chance*, this current has ebbed and flowed over the past 200 years, but it has always been part of our political biosphere. Hanson is just the latest in a long line of chancers, shysters and political hucksters to try their luck at fishing for votes in this disreputable stream.

But what is not true of the One Nation vote may be true of the broader shift to the minor parties. In other words, the centrality of race (and racism) to One Nation is proof neither for nor against the importance of global economic forces in the prolonged structural shift towards the minor parties.

Complementing David's insightful analysis of the forces driving the One Nation vote with a matching analysis of the broader forces behind the shift to the minor parties is an essay-length challenge, currently beyond the limits of this correspondence section, and this correspondent. However, as a starting point I would suggest that just as readers should observe closely the peripheral One Nation voters to get a sense of the major political shift of our time, they should also observe closely the peripheral economic issues now broadening the appeal of One Nation, not just its cancerous racist core.

In her second coming, the White Queen has readily embraced localism, protectionism, economic nationalism and hostility to globalisation (in particular, to skilled migration and free trade) as a useful, if secondary, string to her bow. Pauline Hanson doesn't have the capacity to crystallise these economic discontents into a major national movement. But other, more adroit political leaders, including those currently within the major parties, do have this ability. Some form of protectionism is a recurring theme across almost all of the minor parties, even if it does not dominate One Nation's rhetoric, and it seems to be gaining favour in both major parties at considerable speed.

Marr strips away much of the euphemistic claptrap we use to describe One Nation and calls it like it is. One Nation is a racist political party, with a racist mission, led by a racist leader. Deal with it. However, it is essential that readers do not confuse political vulnerability to the call of racism with the broader shift to minor parties. To do so would risk making two significant political mistakes.

First, we risk failing to perceive an ominous, and imminent, threat to Australia's stable political culture. Second, and more significantly, we risk absolving ourselves of the responsibility to radically reform today's mainstream political offerings in order to re-engage the growing army of voters willing to support Pauline Hanson and populist leaders like her.

I agree with David's conclusion that the White Queen will never get the chance to rule her kingdom, but if we are lulled into a sense of complacency by One Nation's profound political flaws, another fallacious monarch from the Hanson side of the family may well get his or her chance.

Lachlan Harris

John Daley

Pauline Hanson is a mesmerising figure. Her political comeback rivals Lazarus with a triple bypass. Her distinctive opinions fascinate and provide fresh copy for media. She brings an Australian vernacular to the political populism on the rise around the world.

David Marr eloquently captures the highs, lows and more recent highs of Hanson's career. And he uses the Australian Election Study to dissect her support base: Australian-born, trade-educated, opposed to migration.

But this focus on Hanson overlooks a much bigger picture. In the 2016 federal election, minor parties – that is, not counting Labor, the Liberals, the Nationals and the Greens – gained 26 per cent of first preference votes in the Senate, up from 11 per cent in 2004. A good chunk of the electorate is manifestly unhappy and increasingly voting for "anyone but them."

Hanson is benefitting from this rise in support for anyone other than the mainstream parties. But she is a relatively small part of the trend. Nationwide, she picked up only 4 per cent of the Senate vote. Even in her power base of Queensland, Hanson won less than a third of the minor-party vote.

The minor party vote is highly fragmented
First preference Senate vote share, minor parties (not LNP, Labor, Greens), 2016

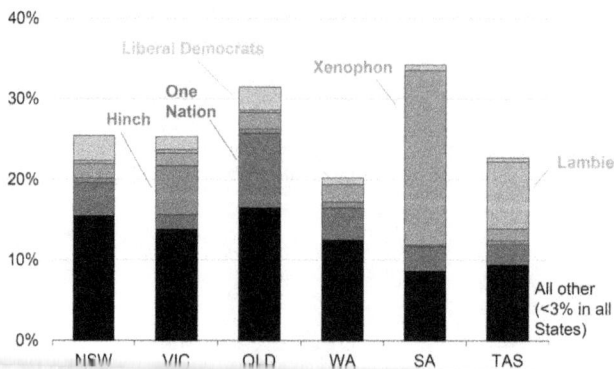

Source: *AEC*

Part of the minor-party vote depends on name recognition. In Victoria the lion's share went to Derryn Hinch, in South Australia to Nick Xenophon, and in Tasmania to Jacqui Lambie. These parties were barely visible outside their "home" state.

Pauline Hanson's One Nation may well implode again. As *The White Queen* documents, the flaws that led to the party's defeat in 1999 may replay two decades later. But the rise of minor parties is likely to continue – it's bigger than Hanson.

What explains this turn away from our established political parties?

Minor parties do well in regional areas, and they're getting stronger. In 2016 minor parties polled roughly 35 per cent of Senate first preference votes in the regions, up from 15 per cent in 2010. This pattern holds across all states: the minor-party vote grows as one travels further away from the state's capital.

Minor party votes are growing, particularly in the regions
1st preference Senate votes to minor parties (not LNP, Labor, Greens)

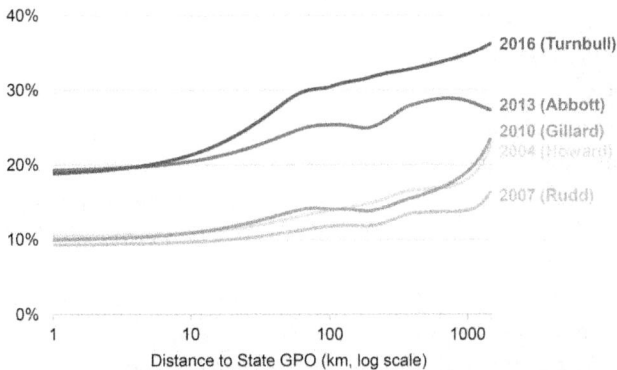

Source: *AEC, Grattan Institute analysis*

Hanson is part of this pattern. Although she has supporters in the cities – and they contain most of the population – her Senate vote in 2016 was strongly regional. She won less than 4 per cent of the vote in the six electorates within twelve kilometres of the Brisbane GPO; she won more than 10 per cent of the vote in eight of the nine electorates more than 100 kilometres from Brisbane (the exception was Leichhardt, which covers Cape York and Cairns).

There is a strong regional element to the electoral appeal of Pauline Hanson

Pauline Hanson One Nation Party first preference Senate vote share by electorate division, Queensland, 2016

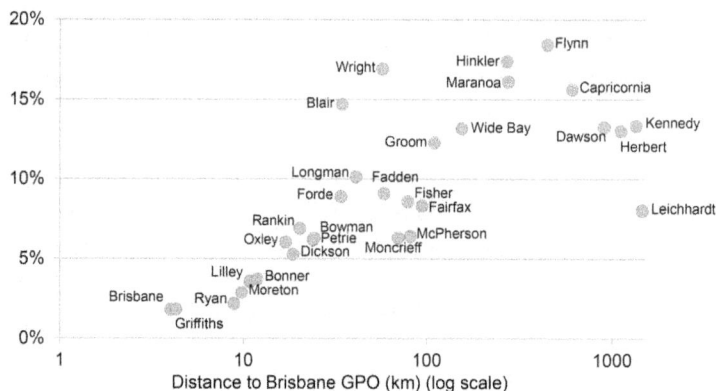

Source: AEC, Grattan Institute analysis

Political discontent in Australia's regions is not new. As Judith Brett showed in her 2011 Quarterly Essay *Fair Share*, regions have long resented their lot in Australian politics. Despite the mining boom, cities are growing faster than regions. Many regional areas are losing population, as regional towns grow only by draining population from their hinterland.

But it's not obvious why these economic trends should translate into different politics now. The past decade simply continues the population trends of the previous century. And income growth per capita has been almost identical in cities and regions over the last decade. In contrast to the United States, those remaining in Australia's regions aren't doing too badly.

Nor is it obvious why race should translate into different politics now. Hanson's resurgence, and the continued rise of other minor parties, does not coincide with any obvious increase in concerns about migration. Marr reproduces the Scanlon Foundation's survey answers on attitudes to multiculturalism and Muslims. Neither have changed materially over the past five years. In the AES, the belief that "immigrants increase crime" has fallen over fifteen years, in both cities and regions. And the belief that "immigrant numbers should be reduced" is well down from its peak in 2010.

It may be that feelings of insecurity – perhaps motivated by sluggish economic growth and the constant discussion of terrorism – have led some people to close ranks and resist the "other" more. Such explanations of outsider politics are gaining currency in the United States and elsewhere, often drawing on Karen Stenner's book *The Authoritarian Dynamic*. People are more likely to have such "authoritarian" attitudes if they particularly value community belonging, and if they have lower levels of education – both more prevalent in regional areas. In these areas, where the share of the Senate vote going to Hanson and other minor parties is highest, migrants are more likely to be seen as the "other" because few migrants live there. Ironically, every time mainstream politicians emphasise the threat of terrorism, they may be promoting the rise of minor parties.

Anti-migrant sentiment is not the only dynamic behind the spectacular rise of minor parties. Even though concerns about migration are much lower in the capital cities, minor parties are also gaining traction there. Many of the growing minor parties such as Derryn Hinch's Justice Party and the Nick Xenophon Team do not play the politics of race that are the core of Hanson's appeal. Voters for minority parties other than One Nation favour migration at least as much as LNP voters.

Instead, many minor parties are appealing primarily to falling trust in government and politicians, as documented by the AES and the Edelman Trust Barometer. As Donald Trump put it, voters want to "drain the swamp." Confidence in institutions is falling even among those comfortable with migration. Around the globe – and in Australia – minor parties vary in their attitudes to migration, to free trade, to size of government, and to inequality. But all of them explicitly claim that establishment politics is no longer serving the public interest. And this claim clearly has electoral appeal.

The biggest challenge to established political parties is not Hanson's politics of race. It is her participation in a much bigger anti-establishment movement that challenges traditional political parties to rebuild voter trust in government. In many countries, this movement has reached the threshold where it can win elections. In a democracy, if governments lose the trust of the electorate, then electorates will ultimately rebuild governments.

John Daley

Ketan Joshi

While reading David Marr's *The White Queen*, I was reminded of an SBS documentary on Hanson that aired in mid-2016. *Pauline Hanson: Please Explain!* featured a brief, candid exchange between Hanson and a dark-skinned crew member standing nearby. Hanson is at ease as she waits for filming to start, and her mind wanders.

"It's my favourite topic. Ref-u-gees. You're not going to tell me you're a refugee, James, are you?"

"No. Aboriginal."

"Really? I wouldn't have picked it. It's good to see that you're actually, you know, taking up this and working."

The sequence is shot like a mockumentary, à la *The Office*, with a quick zoom in and lingering reaction. It's light frivolity. When I tweeted a link to the clip, I expected outrage and re-tweets from the coiled potential energy of social media, but I got nothing. The clip was the same Hanson as in the '90s – casually cruel to dark-skinned Australians – but this time she elicited no public denouncements of racism. Despite the prognostications of right-wing columnists and talking heads who proclaim that "you can't say anything anymore due to political correctness," we've shifted to a place where casual racism, official racism and the foghorn of race-baiting draw muted responses.

I'm terrified at the resurgence of this phenomenon, and how rare it is now to hear someone calling out racism, or even using the term "racist." The fact we've seen it all before is making us complacent – but complacency is fertile ground for hate to grow.

David Marr's *The White Queen* is a brilliant piece of causal surgery, shedding light on how we got here. It delivers something sorely needed in the ugly half of this truly horrible decade: an empirical framework for the contention that racism is a necessary and sufficient factor in the rise of One Nation. Without hatred of ethnic groups, the party would not exist. This seems obvious but, as Marr highlights, referring to a racist movement as racist has been marked as forbidden territory for both major parties.

There were calls in mid-2016 to refrain from labelling Hanson or her voters racist, and to empathise with the economic plight of her impoverished, anxious rural supporters. I read piece after piece urging kindness, discussion and cups of tea. Each of these pleas came from white people. I was furious: even if the argument was right, that economic injustice was the engine of their prejudice, it was no reason to discard the concept of social barriers to racist abuse and adopt a model whereby we politely ask our spittle-flecked abusers to have a chat.

Marr's analysis demolishes the argument that Hanson's supporters are pushed unfairly into their racist malaise by economic change. This cohort is not poor, their jobs aren't threatened, and there's a sizeable contingent hailing from the 'burbs. They have a comfortable perch from which to harm minorities. Much of their sentiment stems from nostalgia for a racially pure country. No one demands of this group that they explain or justify this reversion fantasy, nor are they subject to any social penalties for urging decision-makers to adopt the impossible, at the cost of our nation's prosperity and global standing.

Nor does Hanson herself face any barriers when she encourages those who are hostile to other cultures. This is shown through some truly mean-spirited statements. "It's not about what you want to say, it's the right of being able to have an opinion. Because I can look at someone of another culture ... and I might say I don't particularly like your cultural dance, or I don't particularly like your cultural dress ... but they may be offended by it," declared Hanson, during a debate on section 18C of the *Racial Discrimination Act*.

Hanson's call for the right to attack others' food, culture, dress and dance signals a desire for homogeneity. This is a broad spray that doesn't single out any culture – her recent targets (Muslims) and old foes (Asians) all qualify. But Hanson saves the most potent form of collective suspicion for Muslims: "But you tell me, you line up a number of Muslims, who's the good one? Who's not?" Reducing a complex population of humans to a flat monochrome, whether to instil fear or to erase culture, is horrifying. I read those words and I'm screaming internally. Yet it feels, profoundly, that Hanson has paid no social or political cost for calling for direct attacks on the existence of other cultures in Australia.

Although Marr highlights several disparities between Trump voters and One Nation voters, they share something in common. The recently released 2016 American National Election Study, similar in length and style to the Australian Election Study, found that "since 1988, we've never seen such a clear correspondence between vote choice and racial perceptions." You could use an American citizen's views on immigration to accurately predict whether they'd support Trump, just as you could for Australians and One Nation.

At the time of writing, Turnbull has "scrapped" (renamed) 457 visas for foreign workers and introduced more hurdles for citizenship, including prodding applicants for their adherence to "Australian values" with odd questions blatantly inspired by paranoia about Islam. Hanson claimed responsibility minutes after Turnbull's changes were announced – a repeat of her taking credit for Howard's policies decades ago. Although comparisons between Turnbull's changes and Howard-era policy are numerous, one startlingly significant difference exists: no MPs are standing up in disgust at race-baiting. There are so few influential figures willing to point directly at the racist engine of this movement, whether in reference to the noise emanating from Hanson and her people, or the echoes reflecting back in policy from the Liberal Party and utter silence from the Labor Party.

A Sky News journalist tweeted in late 2016 that an "honest conversation [is] required about ethnicity" after four Australian-born men with Lebanese ancestry were arrested on terror charges. Rowan Dean, on the same network, complained that Victoria ought to feature on its traffic lights "a Sudanese guy with a crowbar flashing up on the lights to warn you that you're about to be carjacked." Immigration minister Peter Dutton linked the ethnicity of terror suspects to immigration policy: "The reality is Malcolm Fraser did make mistakes in bringing some people in the 1970s and we're seeing that today. We need to be honest in having that discussion. There was a mistake made." Hanson has taken to imitating Nazi rhetoric framing Judaism as a disease: "Islam is a disease; we need to vaccinate ourselves against that."

There was a time when pushing the ludicrous concept of an ancestral tendency towards crime would have had immediate social and political costs, in the same way as an MP advocating for policy based on phrenology. Not anymore. We're unsurprised, and we're tired.

There's also a new nervousness around the idea of collective social barriers to harmful behaviour. So much discourse on racism in Australia has focused on legislation – the troubled 18C debate. Those who do call out racism (or the legislative echoes of racism) with volume and intensity are "snowflakes," labelled hypersensitive for standing guard at the social barriers erected to restrict cruelty. Conversely, people who are deeply unsettled by skin colour boast of their hard-headedness and uncompromising grasp of the truth. The unacceptability of racism is a concept that has been eroded by the truly crappy libertarianism imported from America – the same bloody-minded school of thought that gave us climate-change denial and campaigns against plain packaging.

Maybe social outrage isn't the best approach to pushing back against the resurgence of racism. But we're badly underestimating how vulnerable we are to a sharp upswing in publicly supported cruelty. Many millions of Americans watched Donald Trump mock a disabled journalist at a rally, offer to pay the legal fees of anyone assaulting black protesters and brag about physically assaulting women. That they still voted for him is a data point about our species that I will never, ever forget.

I try to temper my fears about living safely and securely in Australia with my experience – it's been almost entirely wonderful. Howard baulked at the idea of multiculturalism, seeing it as a belief that "it is impossible to have an Australian ethos." My personal experience, alongside the statistics laid out in Marr's essay, contradicts that idea. Ancestry, ethnic identity and variations in physicality aren't worthy of panic. Relax. The people telling you to be afraid and anxious are *probably lying and can't be trusted.* Multiculturalism isn't scary.

How long can this majority, too level-headed to feel the hate and fear that Hanson advocates, last without protective social (and, more controversially, legislative) barriers erected around racial discrimination? It's hard to predict, but David Marr's *The White Queen* takes the first step on the road to recovery: describing Pauline Hanson, her party, her people and the governmental echoes of racist anxiety for what they are.

Ketan Joshi

Correspondence

Philip Dorling

Twenty years ago, in September 1996, I was one of the relatively few people present in the House of Representatives chamber when Pauline Hanson gave her infamous first speech in federal parliament. The sole Labor MP in attendance was the Member for Kingsford Smith and Shadow Minister for Foreign Affairs, Laurie Brereton. Laurie had drawn the short straw to be the duty shadow minister present when Hanson was scheduled to speak. As Brereton's foreign affairs adviser, I came down into the chamber with some papers for him to sign, and was seated in the advisers' box when Hanson rose to speak. She was clearly nervous, her voice quavered as she spoke, but there was little that was equivocal in her words. It was a speech that seethed with resentment, anger and undeniable racism. David Marr in *The White Queen* is absolutely right to see white nationalism and racism, the heritage of the old White Australia, as the dark centre of Hanson's worldview.

In that infamous first speech, Hanson's primary focus was on what she saw as "the privileges Aboriginals enjoy over other Australians." She assailed what she claimed to be "reverse racism ... applied to mainstream Australians by those who promote political correctness and those who control the various taxpayer-funded 'industries' that flourish in our society servicing Aboriginals, multiculturalists and a host of other minority groups." These lines still feature regularly in her political repertoire. What generated the most controversy, however, was Hanson's declaration that Australia was "in danger of being swamped by Asians." Asians, she declared "have their own culture and religion, form ghettos and do not assimilate." Unskilled migrants were taking jobs from Australians and could not be trusted to "give this country their full, undivided loyalty." Watching from the floor of the House of Representatives chamber, I could not help but note the Coalition backbenchers – and indeed the government chief whip, who rose to congratulate Hanson warmly after her speech – saying she had said things that needed to be said. My boss Brereton left the chamber shaking his head, describing Hanson as "toxic" and her speech as "the most poisonous thing" he had heard in twenty-five years in state and federal parliaments.

Over the next two years I spent a lot of time working with Brereton on matters relating to Hanson and her new One Nation party. Anxious about her impact on Australia's relations with our Asian neighbours and trading partners, but even more concerned about the impact she was having on community harmony, Brereton favoured Labor taking a hard line against One Nation, but there was a lot of hesitation within the party about confronting her as a racist. Labor leader Kim Beazley and others in the party leadership didn't want to alienate her supporters. Like Prime Minister John Howard, they wanted to dog-whistle to Hanson's supporters. It was no easy decision but eventually Labor decided to put One Nation last on how-to-vote tickets and to press Howard and the Coalition to do likewise. Thanks to that preferencing by both major parties, and her own policy debacles and tactical misjudgments, Hanson hit the political fence at the 1998 federal election. At that time I, and many others, thought she was finished – a conclusion reinforced as her party imploded and lost support. In the aftermath of the 2001 federal election, it looked that Howard's *Tampa* ploy and the political impact of 9/11 had decisively turned the politics of immigration and national security to pull her constituency back to the Coalition.

For a long time that was true and Hanson hung around the margins of national political life, excluded, as Marr notes, in part by the decisions of the major parties, and many minors too, to preference against her. But two decades on, she's back, now in the Senate, accompanied by three colleagues. One Nation has re-emerged, not only on the federal scene but also in state politics, with three upper-house representatives elected to the Western Australian parliament and one member in the Queensland parliament (a Liberal National Party defector). Opinion polls have One Nation tracking consistently at around 10 per cent nationally and significantly higher in Queensland, where the party appears to have good prospects at the forthcoming state election. Twenty years on from Hanson's first parliamentary speech, I find myself doing research for the Australia Institute on the political and policy implications of this nativist, right-wing racist surge. I can't say I anticipated that twenty years ago, or for that matter ten or five years ago.

Marr brilliantly chronicles Hanson's arrival on the national political stage, especially how the door was opened for her by John Howard. He rightly acknowledges her great resilience and determination and her nature as a political animal irresistibly drawn to the rewards of political life. Marr also elegantly uses the fruits of extensive social research to show that Hanson's core constituency never evaporated and has instead remained a persistent strand in national life, not confined to a particular cohort of older Anglo-Australians but being renewed and reborn in both regional Australia and the outer suburbs of our capital cities. Although Marr is right that One Nation would disappear without Hanson, the social research behind *The White Queen* also shows that there is an enduring base of support that would remain, perhaps to be picked up by another right-wing movement and leader.

The White Queen is a masterful historical, socio-political analysis. Marr's insight is less robust when it comes to the future. This is not a criticism; the sheer perversity of human nature makes political punditry a dodgy business at the best of times. That said, however, some wishful thinking is evident in Marr's discussion of the result of the Western Australian state election, which came as he was finishing his manuscript. Marr describes the election as a "fiasco" for One Nation, and one has a clear sense that he would like to see this as a turning of the tide, where questions from ABC journalist Barrie Cassidy showed Hanson "in plain light to be a crackpot."

In large part Marr reflects the instant assessments of media commentators who described One Nation's performance in the Western Australian state election as "a disaster" and an "epic fail." Those commentators focused on One Nation's 4.86 per cent vote across lower-house seats, a figure that was only marginally higher than the party's 2016 Senate vote and well below expectations based on opinion polls. However, the commentators curiously failed to recognise that One Nation did not run candidates in all lower-house seats. In fact, One Nation doubled its support in Western Australia in the seven months between the July 2016 federal election and the March 2017 state election. With voter support rising from just over 4 per cent to more than 8 per cent across the lower-house seats the party contested and in its upper-house vote – only just behind the Greens – reports of a One Nation debacle were greatly exaggerated. Controversy in the final week of the campaign over Hanson's views on vaccinations and Vladimir Putin probably had little impact. One Nation won three upper-house seats, a significant success. There was no debacle, perhaps only a mismanagement of expectations.

With one Western Australian senator, three state legislative councillors, increased personnel and administrative resources, a party office that has run a statewide campaign and significant public funding to reimburse campaign expenditure, One Nation now has considerably strengthened its prospects to retain a Western Australian Senate seat in the next normal "half" Senate election in 2018–19. One Nation may be on track to significantly increase its Senate representation beyond 2019 and through to 2025. The prospect of One Nation securing the balance of power in the federal Senate on a long-term basis should be taken quite seriously. Marr might have usefully looked ahead to the political and policy implications of that prospect.

It is also worth noting that Hanson's place in Australian politics has always had an important international context. Her initial rise was fuelled by an Anglo-Australian backlash against the end of the White Australia policy and the inflow of Asian migration in the 1980s and early '90s. Hanson's declaration that Australia would be "swamped by Asians" had a specific context of Australia shifting away from economic protectionism and embracing the opportunities, and perceived risks, of deep engagement with Asia. The legacy of 9/11 and the persistent threat of radical Islamic terrorism has given Hanson a new and much more potent message, successfully mobilising a new generation of supporters. This is likely to pay political dividends for quite some time. The election of Donald Trump as US president and the success of the "alt-right" in Europe have undoubtedly energised Hanson's supporters and increased the efficacy of her political brand.

But it would be wrong to assume that Hanson will stay solely focused on Muslims and that she thinks "Asians are old hat," as Marr puts it. Although she and her party have gone to considerable lengths to deny that they are "anti-Asian," including recruiting a small number of Chinese-Australian members and nominating a couple as candidates, Hanson has never explicitly repudiated her past anti-Asian rhetoric. In a press conference immediately after her election to the Senate, Hanson publicly reaffirmed her 1996 claim that Australia was at risk of being "swamped by Asians," saying that was a "reality" in Australian cities. "You go and ask a lot of people in Sydney, at Hurstville or some of the other suburbs," she said. "They feel they have been swamped by Asians and, regardless of that now, a lot of Australians feel that Asians are buying up prime agricultural land, housing. You ask people in Melbourne how they feel about it as well."

Speaking in the Senate in September 2016, Hanson went out of her way to revisit her 1996 claims, saying they were made "as a slap in the face to both the Liberal and Labor governments who opened the floodgates to immigration … Their push for globalisation, economic rationalism, free trade and ethnic diversity has seen our country's decline." In February 2017 she declared, "I don't change my tune, whichever way the polls are going. If you look at what I said twenty years ago, it's exactly what I'm saying today."

Questioned about Asian immigration by Barrie Cassidy in March 2017, Hanson ducked and weaved to avoid re-igniting controversy but stubbornly insisted that she had been right in 1996 to warn that Australia was in danger of being swamped by Asians and that such migrants form "ghettos in this country that don't assimilate."

One Nation's policies on radically restricting immigration and opposing multi-culturalism are virtually unchanged from the platform Hanson took to the 1998 federal election. For One Nation, unemployment, dole queues, housing prices and demands on social welfare can all be attributed to immigration, predominantly Asian immigration. Foreign, overwhelmingly Asian, students have been targeted as a cause of high rents and house prices.

One Nation has been careful to keep its anti-Asian sentiments in the background. Targeting Islam and Muslims has proved a more potent political strategy. But Hanson is nothing if not an opportunist and in different circumstances she could easily broaden her range of targets. In an international environment in which significant tension, even possible conflict, between the United States and China cannot be ruled out, Australia's large Chinese community – with its strong commercial and people-to-people links to the People's Republic of China – could easily become a target of highly divisive political campaigning. Senator Hanson has long questioned the loyalty of migrants she does not regard as fully assimilated. She has also long complained about the alleged influence of "ethnic lobbies" on the Liberal and Labor parties. Revelations of significant Chinese influence on Australian politics and the position of the Australian-Chinese community in the context of international tension with China would be targets she would be unlikely to miss. Another successful populist politician, albeit of very different values and temperament, Senator Nick Xenophon, has recently warned of the possibility of international tensions generating grave divisions that could tear at Australia's social harmony. The potential for One Nation to play an opportunistic role in this should not be underestimated.

Hanson and One Nation may rise and fall with the broad fortunes of the alt-right overseas, especially the Trump administration, but they will also do so within Australia's unique circumstances as a multicultural but still predominantly European society on the edge of Asia. In that space it is likely there will continue to be deep wells of insecurity that will fuel One Nation for quite some time to come.

Philip Dorling

THE WHITE QUEEN

Response to Correspondence

David Marr

She's looking ragged. The narrative has shifted. For the best part of a year Pauline Hanson was written up as a mighty challenge to the system. She was taken seriously when she claimed to represent on this continent forces reshaping the politics of the world. This was always over the top. I wrote *The White Queen* not only to name her for what she is – the racist leader of a racist party – but to measure her for size. Hanson's people are driven by much the same nostalgia, scrappy education and white pride that marks the backers of Trump, Brexit and Le Pen. But Hanson is not in their league. She doesn't have their smarts, their cash or their numbers. Why? Because she is operating in a better country. We have our demons – race is one of them – but we don't have America's busted working class and we are more welcoming than Europeans ever have been to new faces in the street. Populist causes over there can, at times, muster nearly half the vote. Pauline Hanson remains Our Lady of 10 Per Cent. There's no reason for politics to kowtow to her.

Western Australia proved the point. That preference deal with the Liberals was a mark of deference her numbers never justified. She got what she wanted – a few key seats in the upper house – but One Nation votes were never going to save the Barnett government. The deal tainted the Liberals, enraged the Nationals and opened deep rifts inside One Nation. The party has now become the story: sacked officials and irate candidates venting on television and complaining to the Australian Electoral Commission, while Hanson issues shrill orders from her home at Scenic Rim, Queensland. When a tape emerged of Ashby brainstorming a scheme to defraud the party's election funding – "there's an opportunity for us to make some money if we play this smart" – Hanson proved unable to counter the wave of scorn that broke over her party. It was a terrible look. I don't resile from the core argument of my essay that any effective response to Pauline Hanson demands we address the racism at the heart of her party, but I'm no longer as disdainful as I was when writing *The White Queen* about the strategy of the major parties to deal with the woman: just show her to be another politician.

Her candidates keep causing her grief. In April, Twitter killed the chances of former Queensland policeman Mark Ellis.

> ELLIS: Hi Australia. My name is Mark Ellis and I am the One Nation candidate for Macalister. Pleased to meet you. Let's open a dialogue.
> MICHAEL HING: hi mark, why are all these people calling you a kidnapper on twitter??
> ELLIS: Because of a 30 year old incident of which I am sorry and have learned a lot. It was not kidnapping though.

Ellis was one of a group of cops who rounded up three Indigenous kids in Fortitude Valley at 4 a.m. and dumped them out in the bush. The idea, police explained, was to give the boys, aged twelve, thirteen and fourteen, "a chance to reflect on their misdemeanours." The police were charged with kidnapping but the charges were dismissed after a judge decided the three boys were willing to take a ride out of town.

> ELLIS: Heaps of other things happened in 1994 other than my kidnapping. NAFTA was signed. Historic.

The candidate's attempts to distract his tormentors didn't work. They came back and back to the Fortitude Valley round-up.

> HING: Woah, sounds pretty full on. What happened??
> ELLIS: I don't want to talk about something so far in the past. I would rather talk about my policies.
> HING: What's your policy on kidnapping?

The candidate withdrew, blaming "leftie media" and "pathetic haters" for forcing him out. Perhaps it didn't help Ellis to post on Facebook a snap of himself giving a Nazi salute to a large swastika he'd just mown in his backyard kikuyu. So far, eight of Hanson's candidates in Queensland have disappeared after one embarrassment or another. God knows who may follow.

My respondents have been kind. They say so often what I wish I'd realised as I was writing. Dennis Glover: "Populism is about unleashing passions and hatred and blame." That's perfect. John Daley: "Ironically, every time mainstream politicians emphasise the threat of terrorism, they may be promoting the rise of minor parties." Absolutely. And what an accolade for Tony Abbott PM. Ketan Joshi: "There was a time when pushing the ludicrous concept of an ancestral tendency towards crime would have had immediate social and political costs, in the same way as an MP advocating for policy based on phrenology. Not anymore." That silence protects One Nation.

Two big themes emerge from the responses. I canvassed the first: that we need to realise the gloom driving people into Hanson's camp is not essentially economic. Glover condemns the old left argument that "all grievances are ultimately economic" and Daley has more figures to show Pauline's people are not those out there doing it toughest. The *Atlantic* on 9 May reported the same economic disconnect at work in America:

> Evidence suggests financially troubled voters in the white working class were more likely to prefer Clinton over Trump. Besides partisan affiliation, it was cultural anxiety – feeling like a stranger in America, supporting the deportation of immigrants, and hesitating about educational investment – that best predicted support for Trump.

It may be time for mainstream politicians to enlist historians to help lure back to their old parties those who have come to feel like strangers in this country. As Glover writes: "We can't bring back the old world of tariffs and White Australia and institutionalised sexual inequalities, and nor should we – but we can try to find new ways of expressing the essence of what made the past different to today."

Lachlan Harris is not so sure that old world is dead and gone. As a political professional to his fingertips, Harris is less appalled by Hanson than the general wreckage caused by "forty-five minor parties" like hers that took three million votes from the majors in 2016. The collapse of the two-party system is the second big theme of the respondents. I barely touched on it in the essay. Harris is right to point out that Hanson could disappear tomorrow without politics in this country returning to the simpler patterns of the past. The break-up experienced in the last decade is profound. Harris calls it "the major political shift of our time." He fears politicians may be willing to turn the clock far back to stop further mass defections: "Some form of protectionism is a recurring theme across almost all of the minor parties ... and it seems to be gaining favour in both major parties at considerable speed."

Meanwhile, Hanson is among us and doing her worst. Since *The White Queen* appeared, Malcolm Turnbull has moved the Coalition a little further into One Nation territory. He has attacked foreign workers and raised even higher English-language barriers to citizenship. Hanson had every right to do what she immediately did: applaud these developments as victories for her cause. Next comes drafting a list of Australian values which every new citizen will have to subscribe to. What an exercise! Poets and historians have been at it for a couple of centuries without reaching agreement. Preachers will fight tooth and nail a list that admits how little we care about religion. The culture wars will break out violently along an extended front unless the list honours only core conservative ideals. Pollsters will be ignored whenever inconvenient. The rhetoric will be wonderful. But it won't be the truth. No one coming to this country will have a chance to sign up to a list of Aussie values that admits we have problems – old problems – with race in Australia, but also reserves of decency on which it might be possible to build far better politics than we have now. Then the ugly instincts and crackpot nostrums of Pauline Hanson would be faced and contested by our political leaders. Who knows: that might restore some regard for politicians we've watched over the last decade do little more than scheme to secure their own survival.

David Marr

Anne Aly is the Labor MP for Cowan. She was formerly a professor at Edith Cowan University, researching counter-terrorism and radicalisation.

John Daley is chief executive of the Grattan Institute, where he has published extensively on public policy issues.

Philip Dorling is a senior researcher with the Australia Institute and a former national affairs correspondent for the *Canberra Times*.

Dennis Glover is a Labor speechwriter and the author of *An Economy Is Not a Society: Winners and Losers in the New Australia*. His first novel, *The Last Man in Europe*, will be published in 2017.

Lachlan Harris is a co-founder and the CEO of One Big Switch, and was senior press secretary to Prime Minister Kevin Rudd.

Ketan Joshi is a Sydney-based writer, specialising in energy, science and technology. He has written for the *Monthly*, the *Guardian* and *Cosmos*.

Anna Krien is the author of *Night Games: Sex, Power and Sport*, *Into the Woods: The Battle for Tasmania's Forests* and Quarterly Essay 45 *Us and Them: On the Importance of Animals*. Her work has been published in the *Monthly*, the *Age*, the *Big Issue*, *Best Australian Essays*, *Best Australian Stories*, *Griffith Review* and *Frankie*. In 2014 she won the William Hill Sports Book of the Year Award in the UK.

David Marr is the author of *Patrick White: A Life*, *The High Price of Heaven* and *Dark Victory* (with Marian Wilkinson). He has written for the *Sydney Morning Herald*, the *Age*, the *Saturday Paper*, the *Guardian* and the *Monthly*, been editor of the *National Times*, a reporter for *Four Corners* and presenter of ABC TV's *Media Watch*. He is the author of six bestselling Quarterly Essays.

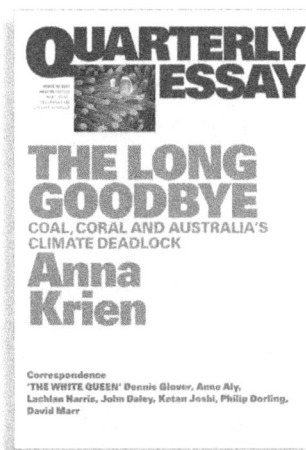

Back Issues: (Prices include GST, postage and handling.)

- ☐ **QE 1** ($15.99) Robert Manne *In Denial*
- ☐ **QE 2** ($15.99) John Birmingham *Appeasing Jakarta*
- ☐ **QE 3** ($15.99) Guy Rundle *The Opportunist*
- ☐ **QE 4** ($15.99) Don Watson *Rabbit Syndrome*
- ☐ **QE 5** ($15.99) Mungo MacCallum *Girt By Sea*
- ☐ **QE 6** ($15.99) John Button *Beyond Belief*
- ☐ **QE 7** ($15.99) John Martinkus *Paradise Betrayed*
- ☐ **QE 8** ($15.99) Amanda Lohrey *Groundswell*
- ☐ **QE 9** ($15.99) Tim Flannery *Beautiful Lies*
- ☐ **QE 10** ($15.99) Gideon Haigh *Bad Company*
- ☐ **QE 11** ($15.99) Germaine Greer *Whitefella Jump Up*
- ☐ **QE 12** ($15.99) David Malouf *Made in England*
- ☐ **QE 13** ($15.99) Robert Manne with David Corlett *Sending Them Home*
- ☐ **QE 14** ($15.99) Paul McGeough *Mission Impossible*
- ☐ **QE 15** ($15.99) Margaret Simons *Latham's World*
- ☐ **QE 16** ($15.99) Raimond Gaita *Breach of Trust*
- ☐ **QE 17** ($15.99) John Hirst *'Kangaroo Court'*
- ☐ **QE 18** ($15.99) Gail Bell *The Worried Well*
- ☐ **QE 19** ($15.99) Judith Brett *Relaxed & Comfortable*
- ☐ **QE 20** ($15.99) John Birmingham *A Time for War*
- ☐ **QE 21** ($15.99) Clive Hamilton *What's Left?*
- ☐ **QE 22** ($15.99) Amanda Lohrey *Voting for Jesus*
- ☐ **QE 23** ($15.99) Inga Clendinnen *The History Question*
- ☐ **QE 24** ($15.99) Robyn Davidson *No Fixed Address*
- ☐ **QE 25** ($15.99) Peter Hartcher *Bipolar Nation*
- ☐ **QE 26** ($15.99) David Marr *His Master's Voice*
- ☐ **QE 27** ($15.99) Ian Lowe *Reaction Time*
- ☐ **QE 28** ($15.99) Judith Brett *Exit Right*
- ☐ **QE 29** ($15.99) Anne Manne *Love & Money*
- ☐ **QE 30** ($15.99) Paul Toohey *Last Drinks*
- ☐ **QE 31** ($15.99) Tim Flannery *Now or Never*
- ☐ **QE 32** ($15.99) Kate Jennings *American Revolution*
- ☐ **QE 33** ($15.99) Guy Pearse *Quarry Vision*
- ☐ **QE 34** ($15.99) Annabel Crabb *Stop at Nothing*
- ☐ **QE 35** ($15.99) Noel Pearson *Radical Hope*
- ☐ **QE 36** ($15.99) Mungo MacCallum *Australian Story*
- ☐ **QE 37** ($15.99) Waleed Aly *What's Right?*
- ☐ **QE 38** ($15.99) David Marr *Power Trip*
- ☐ **QE 39** ($15.99) Hugh White *Power Shift*
- ☐ **QE 40** ($15.99) George Megalogenis *Trivial Pursuit*
- ☐ **QE 41** ($15.99) David Malouf *The Happy Life*
- ☐ **QE 42** ($15.99) Judith Brett *Fair Share*
- ☐ **QE 43** ($15.99) Robert Manne *Bad News*
- ☐ **QE 44** ($15.99) Andrew Charlton *Man-Made World*
- ☐ **QE 45** ($15.99) Anna Krien *Us and Them*
- ☐ **QE 46** ($15.99) Laura Tingle *Great Expectations*
- ☐ **QE 47** ($15.99) David Marr *Political Animal*
- ☐ **QE 48** ($15.99) Tim Flannery *After the Future*
- ☐ **QE 49** ($15.99) Mark Latham *Not Dead Yet*
- ☐ **QE 50** ($15.99) Anna Goldsworthy *Unfinished Business*
- ☐ **QE 51** ($15.99) David Marr *The Prince*
- ☐ **QE 52** ($15.99) Linda Jaivin *Found in Translation*
- ☐ **QE 53** ($15.99) Paul Toohey *That Sinking Feeling*
- ☐ **QE 54** ($15.99) Andrew Charlton *Dragon's Tail*
- ☐ **QE 55** ($15.99) Noel Pearson *A Rightful Place*
- ☐ **QE 56** ($15.99) Guy Rundle *Clivosaurus*
- ☐ **QE 57** ($15.99) Karen Hitchcock *Dear Life*
- ☐ **QE 58** ($22.99) David Kilcullen *Blood Year*
- ☐ **QE 59** ($22.99) David Marr *Faction Man*
- ☐ **QE 60** ($22.99) Laura Tingle *Political Amnesia*
- ☐ **QE 61** ($22.99) George Megalogenis *Balancing Act*
- ☐ **QE 62** ($22.99) James Brown *Firing Line*
- ☐ **QE 63** ($22.99) Don Watson *Enemy Within*
- ☐ **QE 64** ($22.99) Stan Grant *The Australian Dream*
- ☐ **QE 65** ($22.99) David Marr *The White Queen*

☐ I enclose a cheque/money order made out to Schwartz Publishing Pty Ltd.
☐ Please debit my credit card (Mastercard, Visa or Amex accepted).

Card No. ☐☐☐☐ ☐☐☐☐ ☐☐☐☐ ☐☐☐☐

Expiry date /   **CCV**   **Amount $**

Cardholder's name   **Signature**

Name

Address

Email   **Phone**

Post or fax this form to: Quarterly Essay, Reply Paid 90094, Carlton VIC 3053 / Freecall: 1800 077 514
Tel: (03) 9486 0288 / Fax: (03) 9011 6106 / Email: subscribe@quarterlyessay.com
Subscribe online at **www.quarterlyessay.com**

www.ingramcontent.com/pod-product-compliance
Lightning Source LLC
Chambersburg PA
CBHW051336200326
41519CB00026B/7446